Robert Kirchner

On UV-Nanoimprint-Lithography as Direct Patterning Tool for Polymeric Microsystems

TUD*press*

Robert Kirchner

On UV-Nanoimprint-Lithography as Direct Patterning Tool for Polymeric Microsystems

TUD*press*
2012

Bibliografische Information der Deutschen Nationalbibliothek
Die Deutsche Nationalbibliothek verzeichnet diese Publikation in der Deutschen Nationalbibliografie; detaillierte bibliografische Daten sind im Internet über http://dnb.d-nb.de abrufbar.

Bibliographic information published by the Deutsche Nationalbibliothek
The Deutsche Nationalbibliothek lists this publication in the Deutsche Nationalbibliografie; detailed bibliographic data are available in the Internet at
http://dnb.d-nb.de.

ISBN 978-3-942710-65-7

Verlag der Wissenschaften GmbH
Bergstr. 70 | D-01069 Dresden
Tel.: 0351/47 96 97 20 | Fax: 0351/47 96 08 19
http://www.tudpress.de

Gesetzt vom Autor.
Printed in Germany.

TECHNISCHE UNIVERSITÄT DRESDEN

On UV-Nanoimprint-Lithography as Direct Patterning Tool for Polymeric Microsystems

GENEHMIGTE DISSERTATIONSSCHRIFT
zur Erlangung des akademischen Grades eines

Doktoringenieurs
(Dr.-Ing.)

der Fakultät Elektrotechnik und Informationstechnik
der Technischen Universität Dresden

vorgelegt von

Dipl.-Ing. Robert Kirchner
(geb. am 02.09.1981 in Pirna)

Dresden
April 16, 2012

Vorsitzender:	Prof. Dr.-Ing. habil. G. Gerlach	(Technische Universität Dresden)
1. Gutachten:	Prof. Dr.-Ing. habil. W.-J. Fischer	(Technische Universität Dresden)
2. Gutachten:	Prof. Dr.-Ing. H.-C. Scheer	(Bergische Universität Wuppertal)

Tag der Einreichung: 08.08.2011
Tag der Verteidigung: 29.09.2011

Acknowledgements

This work is the result of my PhD research within the DFG Graduiertenkolleg[1] "Nano- and Biotechnologies for Packaging of Electronic Systems" during the last four years since July 2007. All practical laboratory work was done at the Institute for Semiconductors and Microsystems (IHM) at the Technische Universität Dresden and the Fraunhofer-Institute for Photonic Microsystems (IPMS) in Dresden which I owe both great gratitude for all the support during this time.

I would like to express my greatest thanks to *Prof. Dr. W.-J. Fischer* for initiating and for supervising this work. Especially his scientific and organizational support throughout the last four years contributed significantly to this work. I like to express further my deepest gratitude to *Prof. Dr. H.-C. Scheer* for taking the time to co-supervise this work and to discuss many particular results. I would like to thank further her students and staff for taking the time to discuss major parts of this work during the finalization of the manuscript. I would like to thank also deeply *Prof. Dr. J. W. Bartha* for the technological support and access to all the required tools.

Very special thanks go *Dr. H.-G. Braun* for his lectures and advices, support and the time for many questions, which all greatly contributed to the quality of this work. Sincere thanks also to *Dr. B. Adolphi* for the very good collaboration, support and advices during all the time. I further deeply thank *A. Jahn* for giving profound insights into lithography and a lot of support for many new ideas. *Dr. U. Künzelmann*, *Dr. V. Neumann*, and *Dr. A. Türke*, thank you for opening the world of chemistry for an engineer and for in-depth discussions. Many thanks go of course to *S. Waurenschk* for her engagement to organize the chemistry laboratory and for introducing the "magic" of piranha cleaning - a very important groundwork of this thesis. Many thanks go to *L. Nüske* at IPMS for showing the differences between academic research and microsystems manufacturing technology and for keeping the focus on the technology development. Many thanks go also to all the other colleagues at IHM, IPMS and within the research training group for all the small but very important contributions. I cannot mention everybody - but you know who. *R. Landgraf* and *A. Finn* thank you very much for all the fruitful discussions and not only climbing the academic mountains.

Of course, no technological thesis with such a broad spectrum is without the support of many students. First of all I wish to express my deepest thanks to *L. Teng* for his sedulous laboratory work. A lot of thanks go to *A. Kleiner* for her in-depth measurements and investigations and great collaboration. Many thanks also to *M.-K. Kaiser*, *B. Lu*, *N. Tham*, *A. Naumann* and *E. Möller*.

No work over a long time is without the support of many good friends. Greatest thanks to *E. Ohlmer*, *S. Krone* and *E. Barth* and all the other I cannot mention here. However, the biggest thanks for all the support, understanding and patience during the last years is due to my parents, my family and to *K. Wegner* and her family!

* Robert Kirchner (Dresden in February 2012)

[1]Research Training Group

Abstract

For the 21st century, the century of the photon, revolutionary technological innovations are expected for the socially important fields of telecommunications, data transmission, mobility, health-care, lightning, imaging, manufacturing technologies as well as sensing. Innovations will come from further reduced device dimensions and from heterogeneous integration of new device functionalities at the nano-scale into microsystems. Polymers and respective patterning techniques will become a major part of future photonic technologies. Techniques based on nanoimprint lithography are emerging as 3D polymer patterning technique. The current focus of patterning techniques on binary patterning for lithography and CMOS will be complimented by new techniques for the direct replication of devices in a broad spectrum of polymers. Direct patterning requires methods which allow both high aspect ratios and a moderate to high resolution.

This work developed the basic techniques for a UV-assisted nanoimprint manufacturing technology for the direct patterning of a photonic resonator device, a so called microring resonator. A vapor phase based molecular deposition system for perfluorinated silanes as antisticking layers was developed. The achieved silane layers were very dense monolayers with a low surface free energy, a high surface coverage and had a high stability against typical wet mold cleaning procedures and degradation by UV-NIL. A new hydroxyl-functional imprint material was developed as variation of existing imprint polymer formulations. The availability of the integrated hydroxyl groups for later bioconjugation was demonstrated with a specific fluorine tag. This approach demonstrated a route towards the direct patterning of chemical functional polymer devices. Major technology improvements were also achieved for low-cost imprint molds by picking the already known idea of bonded imprint molds. These molds are based on the adhesive bonding of a chip with the imprint structure onto a supporting plate. Important improvements of the chemical long term stability of the used bonding adhesive during wet-chemical cleaning was reached by vapor phase deposition of fluorinated silanes. A new UV-based imprint technology was proposed based on opaque chips bonded to a transparent support. This new technique, termed non-transparent UV-assisted nanoimprint lithography (NT-UV-NIL), allows the application of cost efficient silicon molds and the use of alternative opaque antisticking layers. Certain imprint structures are only achievable in silicon or other opaque mold materials. NT-UV-NIL offers the direct replication of these molds in UV-curing polymers without modifying the established exposure systems with the UV-source above the mold. NT-UV-NIL is well applicable for polymer direct patterning and lift-off processes.

All these achievements were integrated into the proposal of a UV-NIL based manufacturing technology for microring resonators or other directly patterned devices. In contrast to high resolution nanoimprinting, the focus was put on structure heights in the micron-range. The developed UV-NIL technology was based on the "stamp-and-repeat" principle. In comparison to dispensed, layout specific resist volumes and patterns, the use of spin-coated films of UV-sensitive imprint materials was found to be more failure resistant and technologically easier to implement. The pre-exposure of adjacent imprint fields is difficult to avoid for "stamp-and-repeat" techniques and had an influence on adhesion promotion and curing of the UV-sensitive imprint material. Optimizing the imprint field density enabled successful imprinting in spin-coated resist films.

Other patterning techniques are also able to reach the aimed resolution and even the high aspect ratio of this work. However, using functional materials at the same time and the ability to up-scale the technology for 3D structures is best achieved with nanoimprint lithography. Direct patterning is well different from binary lithography and new effects especially regarding polymer flow and mold filling were discussed. The microring resonator was used as technology demonstrator and the proposed technique can be well adapted to other UV-curing material systems and for other directly patterned photonic devices.

In summary, this work contributes to the transfer of nanoimprint lithography techniques from a binary lithography to the direct patterning and microsystems technology approach.

Kurzfassung

Das 21. Jahrhundert gilt als Jahrhundert des Photons. Revolutionäre technologische Innovationen werden vor allem für die gesellschaftlich bedeutenden Bereiche der Telekommunikation, der Datenübertragung, der Mobilität, des Gesundheitswesens, der Beleuchtung, der Bildgebung, der Fertigungstechnologien aber auch der Sensorik erwartet. Innovationen werden sowohl von der zunehmenden Miniaturisierung als auch der Heterointegration neuer Funktionalitäten auf der Nanoebene in Mikrosysteme ausgehen. Polymere und entsprechende Techniken zur Strukturierung dieser werden einen wesentlichen Bestandteil zukünftiger photonischer Technologien ausmachen. Der bisherige Schwerpunkt der Fertigungstechnologien auf hochaufgelöster Lithographie und CMOS-Anwendungen wird durch neue Methoden zur direkten Strukturierung eines breiten Spektrums von Polymeren ergänzt werden. Die Direktstrukturierung von Polymeren erfordert Techniken, welche gleichzeitig hohe Aspektverhältnisse als auch moderat- bis hochaufgelöste Strukturen realisieren können.

In der vorliegenden Arbeit wurden speziell für die Direktstrukturierung photonischer Resonatorstrukturen, sogenannter Mikroringresonatoren, die notwendigen Grundlagen einer Fertigung mittels UV-basierter Nanoimprintlithographie (UV-NIL) geschaffen. Im Speziellen wurde ein System zur gasphasenbasierten Molekularabscheidung von Silanen für Antihaftschichten entwickelt. Die erzeugten Silanschichten waren als dichte Monolayer durch eine geringe freie Oberflächenenergie, einen hohen Bedeckungsgrad sowie eine vergleichsweise hohe Stabilität gegen die im Nanoimprint üblichen nasschemischen Reinigungsprozesse und die Degradation durch UV-NIL gekennzeichnet. Weiterhin wurde ein neuartiges, hydroxyl-funktionales Nanoimprintmaterial als Erweiterung bekannter Resistformulierungen entwickelt. Die Verfügbarkeit der eingebrachten Hydroxylgruppen für eine mögliche Biofunktionalisierung der Polymeroberflächen wurde mit einem spezifischen Fluormarker nachgewiesen. Dieser Ansatz zeigt einen Weg auf, mit Hilfe der Direktstrukturierung und UV-NIL chemisch funktionale Polymerstrukturen für photonische Anwendungen zu realisieren. Im Bereich der kosteneffizienten Herstellung von Nanoimprintstempeln gelang eine deutliche Weiterentwicklung der bereits vereinzelt verwendeten, gebondeten Nanoimprintstempel. Bei diesen wird ein strukturtragender Chip auf einem Grundkörper befestigt. Eine wesentliche Verbesserung der chemischen Langzeitbeständigkeit gegen nasschemische Reinigungsverfahren gelang durch die Gasphasenbeschichtung der verwendeten Adhäsive durch perfluorierte Silane aus der Gasphase. Basierend auf gebondeten Stempeln wurde eine neuartige, UV-NIL-Technik entwickelt. Diese wurde als nicht-transparente UV-basierte Nanoimprinttechnologie (NT-UV-NIL) bezeichnet und erlaubt eine deutliche Kostenreduktion durch die Fertigung von Nanoimprintstempeln aus Silizium. Die NT-UV-NIL ermöglicht weiterhin den Einsatz alternativer, nicht degradierender aber opaker Antihaftschichten. Bestimmte Stempelgeometrien sind nur in Silizium oder anderen Materialien realisierbar, aber nicht in den für UV-NIL typischen Quarzstempeln. NT-UV-NIL erlaubt erstmalig die Abformung derartiger Strukturen in UV-härtende Polymere, ohne dass eine Modifikation der üblichen Nanoimprintsysteme mit einer UV-Quelle über dem Nanoimprintstempel erforderlich ist.

NT-UV-NIL wurde für die Polymer-Direktstrukturierung und Lift-Off-Strukturierung entwickelt und bietet vor allem in diesen Bereichen ein großes Entwicklungspotential.

Die geschaffenen Grundlagentechniken flossen in einen Vorschlag für eine UV-NIL basierte Fertigungstechnologie für polymere Mikroringresonatoren ein. Im Gegensatz zu hochauflösenden Nanoimprinttechniken lag der Fokus auf großen Strukturhöhen im Mikrometerbereich. Die entwickelte UV-NIL-Technologie basiert auf einem "Stamp-and-Repeat"-Prinzip. Im Vergleich zum Dispensen eines optimierten und layoutspezifischen Resistvolumens bzw. -patterns war die Verwendung von aufgeschleuderten UV-empfindlichen Imprintmaterialien deutlich fehlerresistenter und technologisch besser umzusetzen. Das für das "Stamp-and-Repeat" schwer zu vermeidende Vorbelichten benachbarter Imprintfelder hatte Auswirkungen auf die Haftvermittlung sowie die Polymerisation des UV-empfindlichen Imprintmaterials. Durch Optimierung der Belegungsdichte wurde "Stamp-and-Repeat" erfolgreich für aufgeschleuderte Schichten verwendet.

Andere Techniken als UV-NIL sind prinzipiell ebenso in der Lage die angestrebten Strukturgrößen und unter Umständen sogar die benötigten hohen Aspektverhältnisse zu erreichen. Aber vor allem die Verwendung funktionaler Polymere und die Möglichkeit 3D-Strukturen realisieren zu können bleibt im Wesentlichen den Imprinttechniken vorbehalten. Vor allem im Bereich der Grundlagenforschung und der Niedrigvolumenfertigung besitzt Nanoimprint zudem einen großen Kostenvorteil gegenüber anderen Techniken. Mikroringresonatoren wurden als Entwicklungsgrundlage für die erarbeitete Technologie und als Demonstratoren genutzt. Die entwickelte Technologie kann allerdings sehr flexibel auf andere direkt strukturierbare Anwendungen und UV-härtende Materialsysteme übertragen werden.

Zusammenfassend leistet diese Arbeit einen wesentlichen Beitrag zum Transfer der Nanoimprintlithographie von einer reinen Lithographietechnik zu einer Technik für die Polymerdirektstrukturierung. Die Direktstrukturierung unterscheidet sich deutlich von der rein binären Lithographie. In dieser Arbeit werden neue Herausforderungen der Direktstrukturierung vor allem im Bereich des Polymerflusses aufgezeigt sowie Lösungskonzepte vorgeschlagen.

Contents

List of Figures

List of Tables

List of Symbols

Symbol	Dimension	Explanation
a	m	Length of active imprint site (on mold)
a_i^3	nm^3	Average atomic volume of element i
A	Nm^6	Lennard-Jones constant for attractive forces
A	m^2	Surface area
A_0	m^2	Unpatterned mold area
A_1	N	Upper limit of dose-response curve
A_2	N	Lower limit of dose-response curve
A_S	m^2	Sidewall area of patterned mold grooves
$AC_{x,calc}$	at%	Fractional atomic concentration element x (calc.)
$AC_{x,meas}$	at%	Fractional atomic concentration element x (meas.)
AVG	-	Averaged value in general
b	m	Length of imprint feature (on mold)
B	Nm^{12}	Lennard-Jones constant for repulsive forces
c	s^{-1}	Photo-polymerization constant for initiator reduction
$c_{x,i}$	nm^{-3}	Abs. concentration of chemical element x in layer i
c_C, c_F	at%	Atomic concentration of carbon (C), fluorine (F)
$C_{x,i}$	-	XPS-measured atomic conc. element x in layer i
d_i	nm	Thickness layer i
D	J/cm^2	Accumulated UV-dose
D_0	J/cm^2	UV-dose during imprint / imprint dose
D_{50}	-	Half max. effective dose / inflection dose-resp. curve
d_{min}	nm	Minimal resolvable structure width
E_B	eV	Binding energy
E_{kin}	eV	Kinetic energy
E_{Ph}	eV	Photon energy
f_i	-	Surface fraction of material i
F_i	%	Surface coverage of layer i
FD	%	Relative functional density of a chemical group
F_{area}	N	Area proportional force
F_{circ}	N	Circumference proportional force
F_d	N	Demolding force
F_{el}	N	Attractive force due to electrostatic interaction
F_{L-J}	N	Lennard-Jones force
F_{vdW}	N	Attractive force due to van-der-Waals interaction
g	%	Relative amount of bindings in a certain XPS peak
G	µm	Proximity gap (proximity lithography)

h_0	nm	Initial resist layer thickness
h_f	nm	Final residual layer thickness
H_{vap}	J/mol	Enthalpy of evaporation
I	kNs	Impulse (mechanical NIL impact on the resist)
I_{max}	mW/cm^2	Maximal intensity (surface power density)
I_t	mol/l	Photoinitiator concentration at time t
I_x	s^{-1}·sr^{-1}	Photo-electron intensity of chemical element x
k	-	Process factor in projection lithography
k_I	s^{-1}	Reaction coefficient for photoinitiation
m	-	Number of features on imprint mold
M	kg/mol	Molecular weight
n	-	Amount of atoms in a layer
n_i	-	Refractive index of material i
N_A	mol^{-1}	Avogadro constant (6.0221367·10^{23} mol^{-1})
N_i	nm^{-3}	Atomic density of a chemical element in layer i
NA	-	Numerical aperture
p	Pa	Imprint pressure
p	cm^2/J	Slope of dose-response curve
p_i	Pa	Generic pressure i
p_{bp}	Pa	Vapor pressure at the boiling point
P_x	nm/(s·sr)	Normalized intensity of chemical element x
r	m	Distance over which attractive force interact
r_l	nm	Lateral etch distance
$r_{m,x}$	-	Measured ratio of elemental concentrations of the chemical element or group x
r_o	nm	Orthogonal etch distance
$r_{t,x}$	-	Theoretical ratio of elemental concentrations of the chemical element or group x
R	J / (K·mol)	Universal gas constant (8.314510 J·K^{-1}·mol^{-1})
R_{init}	s^{-1}	Rate of photoinitiation
R_q	nm	Root mean square surface micro-roughness
s	m	Distance the polymer is squeezed to the next cavity
S_x	-	Standard XPS, single-element sensitivity factor
t	nm	Etch depth of imprint mold
t_f	s	Filling time / time to reach h_f
T	-	Combined transmission and detector efficiency
T_i	°C, K	Generic temperature i
W_a	mJ/m^2	Thermodynamic work of adhesion
W_c	mJ/m^2	Thermodynamic work of cohesion
Y	%	Imprint yield
Y'	%	Corrected imprint yield
z_x	nm	Integration variable for layer x
Z	-	Atomic number

α	dB/cm	Attenuation loss
ϵ_i	-	Volume fraction of material i
γ_i	nm	Inelastic mean free path of electrons in layer i
γ, γ_i	mN/m, mJ/m^2	Surface free energy / surface tension (of component i)
γ_{ij}	mN/m, mJ/m^2	Interfacial energy / tension between surfaces i and j
γ_i^p	mN/m, mJ/m^2	Polar contribution to γ_i
γ_i^d	mN/m, mJ/m^2	Dispersive contribution to γ_i
γ_i^{LW}	mN/m, mJ/m^2	Lifshitz-van der Waals contribution to γ_i
γ^i	mN/m, mJ/m^2	Polar dipole-induced dipole contribution to γ
γ_i^{AB}	mN/m, mJ/m^2	Lewis acid-base contribution to γ_i
γ_i^-	mN/m, mJ/m^2	Lewis-acid contribution to γ^{AB}
γ_i^+	mN/m, mJ/m^2	Lewis-base contribution to γ^{AB}
η_0	Pa·s	Zero shear viscosity
λ	nm	Wavelength
$\lambda_{x,i}$	nm	Inelastic mean free path element x in layer i (approximated by element specific attenuation length element x in layer i)
π	mN/m, mJ/m^2	Spreading pressure
θ	°	Detector angle
θ_c	°	Critical angle (geometric optics)
$\theta_{i,n}$	°	Angle of incidence for material n (geometric optics)
$\theta_{r,n}$	°	Angle of refraction and reflection (different interface) for material n (geometric optics)
Θ_e	°	Effective contact angle (Cassie law)
Θ_i	°	Contact angle on surface / material i
ρ	g/cm^3	Density
σ	-	General standard deviation
σ_x	b (=10^{-28} m^2)	Cross-section photoelectron emission element x
τ	-	Time constant for photo-initiator conc. reduction
ϕ	R [$=1/(\mathrm{m}^2 \cdot \mathrm{s} \cdot \mathrm{sr})$]	X-ray flux
Φ	-	Good-Girifalco thermodynamic parameter for W_a/W_c
Φ	-	Quantum yield of photoinitiators
Φ'	-	Initiator molecules dissociated per absorbed photon
Φ_S	eV	Spectrometer specific work function
Ω	-	Aspect ratio (structure height to structure width)

List of Abbreviations

2.5d	2.5-dimensional
3d	3-dimensional
AFM	Atomic force microscopy
AR-XPS	Angle-resolved XPS
ASL	Antisticking layer
BIM	Bonded nanoimprint mold
bp	Boiling point
CD-XPS	Chemical derivatized XPS
CNP	Combined nanoimprint-and-photolithography
CVD	Chemical vapor (phase) deposition
DOD	Drop-on-demand dispensing
EBL	Electron beam lithography
EGDMA	Ethylene glycol dimethacrylate
FDTS	1H,1H,2H,2H-perfluorodecyltrichlorosilane
FOTS	1H,1H,2H,2H-perfluorooctyltrichlorosilane
HBA	4-Hydroxy butyl acrylate
HMPP	2-Hydroxy-2-methyl-propiophenone
HNM	Hybrid nanoimprint mold
IEP	Isoelectric point
ITRS	International Technology Roadmap for Semiconductors
LDW	Laser direct writing
LW	Lifshitz-van-der-Waals
MRR	Microring resonator
NIL	Nanoimprint lithography
PC	Polycarbonate
PDMS	Polydimethylsiloxane
PEG	Polyethyleneglycol
PFP	Pentafluoropropane
PFPE	Perfluoropolyether
PMMA	Polymethylmethacrylate
PS	Polystyrene
PVC	Polyvinyl chloride
PVDF	Polyvinylidene fluoride
PVF	Polyvinyl fluoride
RIE	Reactive ion etching
RLT	Residual layer thickness
RMS	Root mean square

SEM	Scanning electron microscopy
SFE	Surface free energy
SR-NIL	Stamp-and-repeat NIL
SR-UV-NIL	Stamp-and-repeat UV-NIL
T-NIL	Thermal-based NIL
tBA	*tert*-butyl acrylate
TCSP-MA	3-(trichlorosilyl)propyl methacrylate
TES	Triethoxysilane
TFAA	Trifluoroacetic anhydride
TMS	Trimethoxysilane
UV-NIL	UV-assisted NIL
UVL	Ultra-violet photolithography
vOCG	van Oss, Chaudhury, Good
XPS	X-ray photoelectron spectroscopy

1 Introduction

"640K should be enough for anyone."

(Bill Gates)

Just about a quarter of a century ago, electronic journals and electronic publishing started to revolutionize the scientific community. The knowledge of this world-wide community is available online at any time. Research results are published in increasingly short times as online-first articles. The complete publishing is computerized to a large extent to handle the enormous amount of manuscripts and data. Increasing numbers of research journals are founded. Regardless of individual opinions on the advantages of this development, this trend became only possible due to the continuous improvements in microelectronics and data storage technologies. This is only one example how the performance and the development of the modern world depends on technological innovations.

The development of the "high-tech society" of today depends mainly on two simple principles: i) decrease the size that a certain function requires in a device and ii) add new functions to a device [1]. If possible, new functions should not increase the size of the device. This fact drives the shrinking of the already existing functions. New functions have to be integrated to microelectronic devices from non-electrical domains like fluidics to bridge the gap between electronics and biology. Current research is aimed at understanding and exploiting effects on the nano-scale. These effects must be integrated to and controlled from a macro-sized environment. Current high-tech microelectronics production reached already the 32 nm technology node.

Manufacturing of structures at the nano-scale and a system integration with other functionalities demand dedicated lithography process. Besides lithography techniques which have just a very high resolution, techniques are required that can realize complex and 3-dimensional structures. Even more, they have to be also applicable for new functional materials and have to span multiple scales from nano-sized to micron-sized structures. After all, the technique must be first affordable for research to develop new functionalities and later for production. Many different techniques based on and related to nanoimprint lithography are very promising for individual and application specific solutions.

1.1 Nanoimprint Lithography

Nanoimprint lithography (NIL) is a mechanical patterning technique. It is based on embossing or molding principles that are well known from other industrial fields, but for structures in the mm-range. *Chou et al.* demonstrated in 1995 for the first time a sub-25 nm pattern transfer with a hot embossing technique [2]. This publication was the starting point for an intensive research on NIL. Primarily, NIL was seen as lithography technique for microelectronics. Later, it started to emerge as patterning technique for microsystems. For example, *Chao et al.* used a hot embossing or thermal based NIL (T-NIL) technique to replicate planar optical waveguides [3]. Two basic and quite different NIL-techniques have been established. UV-based NIL (UV-NIL) is based on low-molecular weight, UV-curing compounds, room-temperature processes and moderate pressures. Initial work was for example done by *Haisma* [4] and *Colburn* [5]. T-NIL in contrast uses thermoplastic material and high temperatures and pressures. NIL is partially related the modern manufacturing of optical storage mediums like CDs or DVDs, where a polymer melt is injected into a mastering tool.

NIL requires a master tool or imprint mold with a surface topography (see Fig. 1.1a). The topography of this imprint mold is mechanically transferred into an imprint material being mostly a polymer (see Fig. 1.1b). After detaching the mold, the imprint material represents a surface topography, which is the negative of the master relief (see Fig. 1.1c). The generated height contrast can be binary, 2.5-dimensional[1] (2.5d) or truly 3-dimensional (3d).

Figure 1.1: *Schematics of the process flow of nanoimprint lithography with a) placement of the mold above the imprint material, b) transferring the pattern into a flowable imprint material and c) demolding of the replicated surface topography.*

Binary patterns are required as masks for a structure transfer to the underlying substrate, e.g., by etching the substrate or by a lift-off of deposited materials. Since 2003, nanoimprinting is part of the International Technology Roadmap for Semiconductors (ITRS) [6] and considered as a possible contender for binary pattern replication by optical lithography, which is expected to reach someday its physical limits.

[1] 2.5d is here understood as 2d, or binary, structures being only extended in its height.

Today, NIL gains increasingly interest for 2.5d and 3d relief replication using its unique characteristic as a dedicated polymer patterning technique. 2.5d structures of a sufficient thickness or 3d structures are the basis for functional devices. NIL can create such structures in a single imprint step. This approach of patterning structures for a direct use, being not pattern transfer to a substrate, is shortly termed "direct patterning" throughout this work. Direct patterning is further understood as an alternative to binary mask patterns for polymer devices. Complex, additional pattern transfer processes are not required. One advantage of using polymers for microsystems is the wide spectrum of usable and tunable material properties. For example, polymer channels can be used to confine fluids. Such fluidic channels have been nanoimprinted in the size of DNA, which allows a manipulation of the DNA-elongation at the nano-scale [7]. For such channels only a high resolution is required. Pattern transfer and residual layer issues are not required to be considered. This reduces the patterning process to the basic strengths of NIL being surface topography replication. Polymer photonics was soon discovered as emerging field for nanoimprinting [8] and are expected to benefit currently the most from NIL techniques [9]. Polymeric lenses [10, 11], optical gratings, spectrometers [12] and planar optical waveguides [3] are just a few examples for directly patterned photonic structures.

Nanoimprinting is currently a high-resolution, top-down patterning approach for sub-50 nm binary features and sub-300 nm directly imprinted features. The resolution for directly imprinted features is often limited and one order of magnitude higher because directly imprinted structures often require a higher aspect ratio (structure height to width) than binary patterns. High aspect ratios of 15 were reached [13]. Aspect ratios of 50 were reached for imprinting anodized alumina into metallic glasses and dissolving the mold [14]. However, the molds used bottom-up grown patterns, sometimes with easy to demold sloped sidewalls and high defect tolerances or required mold destruction for structure isolation. Imprintable aspect ratios of top-down patterned, reusable molds are rather in the range of five to ten.

The relevance of NIL must be discussed in the certain field of application. In basic research, NIL offers new understandings of materials and natural sciences. NIL is a simple and cost-efficient way to high resolution lithography for basic research and for understanding nano-scale effects. In this field, NIL is basically a research vehicle. For production, new aspects as yield, defects and process controllability must be considered. However, the NIL resolution for both binary and direct patterning is well below the resolution limit of i-line lithography. The latter is often the affordable limit for microsystems applications, making NIL also interesting for production in this field. The ability to use new materials, topographies and substrates are most important for industrial application of NIL. Industrial NIL is best suited for specialized applications.

After all, NIL is a "one-to-one" lithography technique and the master structures have to be made in the same dimensions as the desired final structures. Even more challenging is that complex 3d patterns have to be mastered with very sophisticated techniques before they can be replicated by nanoimprinting. But once realized as a master structure, nanoimprinting can replicate these patterns multiple times in a very broad spectrum of materials on very diverse and for other lithography challenging substrates like highly reflective or flexible materials. Also an area enlargement for wafer-scale molds can be realized by NIL itself using stamp-and-repeat techniques. Thus, NIL concentrates the effort of advanced lithography on a small (mold) area.

1.2 Direct-Patterning by Nanoimprinting

Optics Photonics are seen as a key enabling technology for the 21st century [15]. Polymer waveguides are an important component of integrated optics. In 2001, polymer waveguides made up only 4.5 % of the integrated optics market. For 2005, the market share reached already 7.9 % [16]. This is equal to an annual growth of more than 14 %. Advantages of polymer waveguides compared to inorganic waveguides made from materials like silicon, silica, gallium arsenide, indium phosphide and lithium niobate are the following [16].

- The single-mode waveguide dimensions (micron range) are larger due to the lower refractive index of polymers.
- Polymeric waveguides have less stringent requirements on the surface roughness due to a lower refractive index.
- Polymers enable a rapid and easy processing.
- Polymer techniques are generally low-cost approaches.
- Polymers enable the integration and tuning of complex optical, chemical and mechanical properties to realize functional devices.
- Polymers allow a device integration on many different substrates due to there compatibility to many different rigid and flexible as well as inorganic and organic materials.

A representative example of the integration of functionalities at the sub-micron and nano-scale into systems at the micron-scale is the so called microring resonator (MRR) (see Fig. 1.2). This resonator requires micron-scale polymer waveguides to confine and transport light from the macro-world to the resonator unit. The resonator unit is the most important functional part of the device and is based on a sub-300 nm gap. Thus, aspect ratios of five to ten are required to efficiently couple light from a straight transport waveguide into a ring resonator waveguide and back again. By designing the ring geometry and coupling distance, frequency selective devices can be built. This can be used for static filters. If the effective refractive index of the resonator can be modified, sensors and active optical devices can be realized.

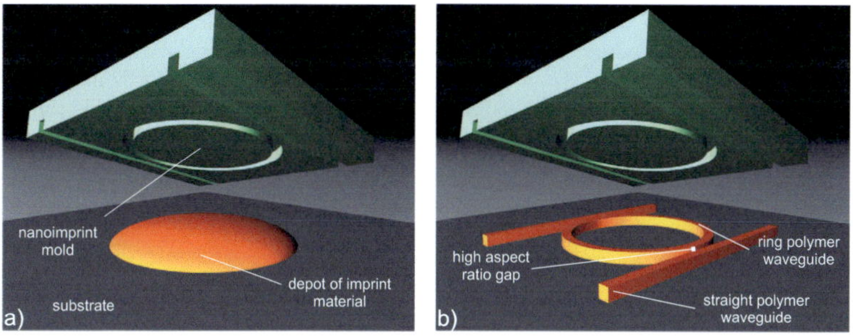

Figure 1.2: *Schematic of a) direct nanoimprinting of a polymer microring resonator into a liquid imprint material depot and b) demolding of the final structure after solidification of the imprint material.*

UV-assisted nanoimprinting techniques Direct UV-assisted NIL (UV-NIL) uses low-viscous monomers as imprint material and UV-light to polymerize the monomers and to preserve the mold structure in the imprint material. The direct nanoimprinting of an MRR was already demonstrated by T-NIL [3, 17]. Even though T-NIL offers some interesting features as post-imprint smoothening of the waveguides, it has some important drawbacks. Regarding these drawbacks, UV-NIL is characterized by

- faster process cycles and less mechanical impact (stress, deformation) on the substrate material due to lower imprint pressures and process temperatures,
- a less complex polymer flow due to the lower viscosity,
- a more efficient synthesis of functional imprint materials and
- the availability of an efficient residual layer free imprinting approach, which preserves functional structures by avoiding post-imprint treatments,

compared to T-NIL. Environmental instabilities of thermoplastics were mentioned by *Kim* [18]. They should not pose a major issue for an MRR used as a disposable part of a biosensor. However, as soon as long term applications are of interest, these instabilities should be kept in mind and cross-linked, UV-NIL polymer should be considered.

Technology Challenges and Current Trends Currently, the most important issues for binary as well as direct nanoimprinting are

- particles and defects,
- technology match with other lithography and semiconductor processing techniques,
- antisticking layer stability and interaction with the imprint material,

- high aspect ratio patterns,
- patterning of functional imprint materials,
- 3d mold manufacturing and replication.

Particles have always been the greatest challenge for nanoimprinting. Particles on imprint molds and substrates originate from standard handling even under clean room conditions. Additionally, sticking resist residues on the mold come from the mechanical contact of the unprotected mold with the imprint material. This is especially critical for high aspect ratio patterns where sticking is more probable. Particles disturb the intimate contact between mold and substrate and can damage the mold. To minimize the risk of particle generation, a well tuned process with a high adhesion of the imprint material to the substrate and a low adhesion to the mold is required. A self-cleaning effect of NIL-molds was demonstrated [19] but is not always sufficient. Currently, soft polymer molds are proposed as an alternative to rigid molds allowing a much more conformal mold contact to the substrate even in the presence of particles and wafer-scale polymeric molds [20].

Due to its mechanical character, the overlay accuracy of NIL is normally in the order of 100-250 nm and thus much lower than the accuracy for optical projection lithography. Using NIL for the first pattern layer, a layer without a high overlay accuracy or for a single-layer approach is the best solution to integrate and match NIL with other techniques. Direct patterning of waveguides and planar optical structures is such a typical single-layer approach.

A continuous degradation of the mainly silane-based antisticking layers (ASL), especially during (UV-NIL) became well accepted in the recent years. Reasons are still object of current research, but the reactivity of resist components with the silane layers is known to be the most important aspect. Less reactive antisticking measures due to modified antisticking layers and resists are required.

High aspect ratios are the characteristic challenge for direct patterning and a well tuned, high adhesion to the substrate is required. The deeper mold structures in direct patterning hamper the mold cavity filling process and the demolding. This is a new field of research in NIL and is maybe even more complex than for binary patterning and low aspect ratio mask structures.

Patterning functional imprint materials and replicating multi-level, 3d structures is the future challenge of NIL. The 3d-mold manufacturing with sidewall profiles and geometries suited for imprinting is probably one of the key-issues for a successful implementation of NIL as future lithography technique. New functional materials also influence the interaction of the imprint material with the ASL on the molds and the adhesion to the substrate. In contrast to binary patterning, these interactions are unique for each new functional imprint material and require, hence, a dedicated technology development.

Particle issues, the technology match with other lithography techniques and true 3d structures are not addressed explicitly in this work. The focus was rather on antisticking layers, functional imprint materials and direct patterning.

1.3 Scope and Structure of this Work

This work was intended to demonstrate a possible route towards direct UV-assisted imprinting of functional materials and to contribute to the growing field of direct patterning for micro-nano-systems. As technology demonstrator a microring resonator was used. To achieve this goal, this work had first established the basic components for an imprint technology. Based on those components, finally a technology capable for a medium-volume production of microring resonators was proposed. This goes well beyond already published work and pure feasibility demonstration of imprinted microring resonators. The document is composed of two basic and two experimental chapters with topical sections on the basic technology components and the final approach towards an imprint technology.

Adhesion and the engineering of it was a major part of this work. Therefore, Chapter 2 discusses in detail the physico-chemical background of adhesion. As an important issue, the forecasting of adhesion is discussed. This is most important for the application of new antisticking mold measures. In particular, the basics for the use of metal-oxides as antisticking layers are discussed. Silane monolayers were covered in a dedicated section. Special focus was put on their assembling and characterization. Chapter 3 deals with planar optical waveguides, lithography methods used to realize them. The focus was put on nanoimprinting in comparison with possible alternative techniques.

In the experimental chapters, results of the relevant technology components are discussed separately. First, the new implementation of a molecular or chemical vapor-phase deposition technique is covered in Section 4.1. The newly developed setup and the resulting antisticking layers were characterized and yielded very stable layers compared to published literature values. The established technique is able to yield conformal antisticking layers equally on very small structures and very high aspect ratio features. To demonstrate the big advantage of UV-NIL in using dedicated and customized resists, Section 4.2 demonstrates in depth the development of a new hydroxy-functional imprint resist. Furthermore, the superficial density of functional groups was varied. Additionally, the ability to bind specific molecules to the chemical functional polymer surface is demonstrated. In Section 4.3, the already existing concept of bonded imprint molds was improved and discussed for a broad spectrum of materials. A real innovation was possible by coating the adhesive with fluorinated silane molecules. This enabled to render adhesives much more stable against wet chemicals like acids. As a breakthrough and major contribution towards the universal application of silicon imprint molds a new technique was proposed, which uses opaque molds, opaque substrates and a UV-exposure. Details of this innovative technique are discussed in Section 4.4. This new technique enjoys a high synergy with the particularly developed technique for bonded imprint molds. Furthermore, it reduces strongly the effort to realize the required waveguide dimensions in a mold because high fidelity etching in silicon is much less challenging than in fused silica. Finally, all basic components are combined to a UV-NIL stamp-and-repeat manufacturing approach of microring resonators in Chapter 5. In contrast to established techniques, this work proposes a stamp-and-repeat technique with spin-coated films. For the first time important issues of the pre-exposure effects on adjacent fields are discussed. The technology is based on a sample resist system. It can, however, be easily transferred to other UV-NIL resist systems for an improved performance of the aimed microring resonator system, which design parameters are developed in an accompanying PhD work [21].

Finally, all achievements of this work are summarized in Chapter 6 and the final conclusion on the usability of UV-NIL for the replication of microring resonators in a medium-volume production is drawn.

Basics

2 Basics of Polymer Adhesion and Silane Monolayers

> "Science has 'explained' nothing; the more we know the more fantastic the world becomes and the profounder the surrounding darkness."
>
> *(Aldous Huxley)*

NIL is based on a mechanical contact between the mold and the imprint material. Engineering and understanding the adhesion of the imprint material to both the substrate and the mold is the key issue for successful, defect-free imprinting results. This is especially crucial for a direct patterning approach using high aspect ratios.

The surface free energy (SFE) is a well established surface characterizing parameter but its applicability to the characterization of antisticking layers (ASLs) is limited. SFE can neither characterize the reactivity of the ASL towards the imprint material nor the stability during the imprint process. The SFE cannot be used to calculate universally the adhesion between the ASL and the imprint material, but it can be used to predict it.

Typical high quality and conformal NIL antisticking layers are prepared by vapor phase silane deposition methods. These silanes have a limited stability during UV-NIL. The most probable degradation mechanism is fluorine abstraction as low molecular weight, volatile compounds. Alternative more stable ASLs are metal oxides or other coatings with low reactivity towards the imprint resist. Metals benefit from mature and well controllable deposition processes in microelectronics. To circumvent problems arising from the opaqueness of metals, NT-UV-NIL or other modifications of the standard "through-the-mold" exposure setup are required.

2.1 Intermolecular Forces

Adhesion Adhesion describes interactions between different surfaces. It results in a certain amount of work required to separate these surfaces after they had been brought into contact [22]. In a classical context, the surfaces are termed adherends. Adhesion is a measure how well a mechanical load can be transferred from one adherend to another. Adhesion can be mediated by a respective bonding material, the so called adhesive [23]. In NIL, the stress should be transferred to the substrate but not to the mold. In other words, the imprint material should adhere well to the substrate but not to the mold. Several forces and mechanisms are the source to adhesion. They can be grouped into physical, chemical and mechanical interactions [23, 24]. The following overview is based on [23] and [25].

Physical Interactions Physical interactions comprise electrostatic forces. The van-der-Waals interactions are a sub-group of electrostatic forces. Electrostatic forces arise from the coulomb interaction of charged bonding partners. Electrostatic forces are important for ionic bonds reaching strengths of more than 100 kcal/mol, which make them one of the strongest interaction besides covalent bonds (see Tab. 2.1). Electrostatic forces are long-range forces with an inverse quadratic relation between the force F_{el} and the distance r between the charged partners.

$$F_{el} \propto \frac{1}{r^2}. \tag{2.1}$$

Van-der-Waals forces can be sub-grouped into dipole-dipole, dipole-induced dipole and dispersion forces. Dipole-dipole interactions result from a shift of partial electric charges within a molecule due to different electronegativity values of chemical elements. For example in the molecule $CF_3{-}CH_3$, electrons are shifted towards the fluorine leading to a permanent dipole molecule [23]. Such dipoles can interact with each other within homogeneous materials as well as with dipoles of other materials, resulting in a short-range attraction of these materials. Hydrogen bonds represent a special type of dipole-dipole interaction. A hydrogen atom of one molecule forms a bond with a highly electronegative atom like oxygen of another molecule. Dipoles can also influence molecules with a symmetrical charge distribution (i.e. non-dipoles). The dipole exerts a force on the electron distribution in the symmetrical molecule, leading to an induced dipole. The result is a net attraction between the primary dipole and the induced dipole. Additionally, also an interaction between molecules of a symmetrical charge distribution is observed. The reason are fluctuations of the electron density within the molecules. This leads to a fluctuating dipole which can interact with other fluctuating dipoles. These forces exist between all materials and are known as dispersion forces. All van-der-Waals forces F_{vdW} are short range forces with the relation

$$F_{vdW} \propto \frac{1}{r^7}. \tag{2.2}$$

For NIL it is important to understand that van-der-Waals forces are always present between mold and resist. It is not possible to eliminate these attractive forces.

Chemical Interactions This are all those interactions that lead to chemical bonds and are distinguished from physical bonds that originate only from attractive forces as discussed above. Chemical bonds can be of ionic, metallic or covalent nature. In an ionic bond electrons are completely transferred from one bonding partner to the other and the bond is based on coulombic attraction between the differently charged bonding partners. Metallic bindings are based on the interaction of delocalized electrons and positively charged metal ions. For NIL, ionic and metallic bindings are not relevant. In covalent bonds, shared electron pairs between bonding partners lead to chemical bonds. The bond strength depend on the bonding partners. An important group of covalent bonds are acid-base interactions of an electron donor providing electrons to an electron acceptor.

Attraction and Repulsion All attractive forces are counterbalanced by repulsive forces due to equivalent charges of the electron clouds of the atoms approaching each other. However, repulsive forces are extremely short-range forces. The relation between attraction and repulsion can be described with the Lennard-Jones theory using the constants A and B to scale the attractive and repulsive forces, respectively. Attractive forces are by definition negatively signed. F_{L-J} is the effective Lennard-Jones force between two binding partners.

$$F_{L-J} = -\frac{6A}{r^7} + \frac{12B}{r^{13}} \tag{2.3}$$

Physical interactions are thermo-dynamically reversible in contrast to chemical interactions. This is, a physically adsorbed molecule can be removed if the adsorption of another molecule or to another surface is energetically favored. Chemical bonds in contrast require a higher energy input to break a bond even if another binding would be more favorable. To have a high and environmentally stable adhesion, chemical adsorption is required.

Table 2.1: *Overview on binding energies of different intermolecular forces that are responsible for adhesion [25].*

Intermolecular force	Binding energy [kcal/mol]
Chemical bonds	
Ionic	140-250
Covalent	15-170
Metallic	27-83
Physical bonds	
Hydrogen bonds	<12
Dispersion forces	<10
Dipole-dipole forces	<5
Dipole-induced forces	<0.5

Mechanical Interactions Adhesion due to mechanical interactions is based on interlocking and friction [24]. Mechanical adhesion is closely related to surface roughness and will be further discussed together with the adhesion models (see Sec. 2.4).

2.2 Surface Energy, Work of Cohesion, Work of Adhesion

The following section is based on [23] and summarizes NIL relevant theories of adhesion. The above mentioned intermolecular forces are responsible for the cohesion of bulk materials. A surface is a discontinuity of the bulk material. Superficial atoms are lacking a similar bulk atom outside the bulk. These atoms are said to have unbalanced forces. These forces are able to do work and thus a certain amount of energy can be ascribed to surfaces. In liquids, this amount of "freely available energy" γ is known as surface tension and in solids as surface free energy (SFE) .

Separating a homogeneous material into two parts requires a certain amount of energy. Two new surfaces are created having each the SFE of the original material. The required amount of energy required for separation is expressed by the work of cohesion W_c and equals two times the SFE γ of the original material.

$$W_c = 2\gamma \tag{2.4}$$

Similarly, the work of adhesion W_a can be defined. It is a measure for the amount of energy required to separate two materials at their interface. Balancing the situation before and after separation gives the well known *Dupré* equation for W_a (see Eq. (2.5)). Before applying W_a, only an interfacial energy is present due to the fact, that the both materials does not balance their bulk forces equally. If they would do so, the interfacial energy would be zero and a homogeneous material would be present. During the separation two new surfaces are created, namely that of material 1 and material 2 each having its own SFE γ_1 and γ_2. γ_{12} is the interfacial energy between both materials. In addition, the system has lost thermodynamic work, namely W_a, that was required for separation.

$$W_a = \gamma_1 + \gamma_2 - \gamma_{12} \tag{2.5}$$

2.3 Measurement of Surface Energies

The surface energy of solids is difficult to access. The most common indirect technique therefore is contact angle measurements. For a direct measurement technique of the SFE as well as its theoretical derivation from basic thermodynamic principles the reader is referred to [22, 23].

The idea of contact angle measurements is to use probe liquids with known surface tensions and to measure their interaction with the investigated solid. Depending on the surface tension of the liquid and the SFE of the surface, the liquid drop will spread on the test surface to a certain extent. Prerequisites for successful measurements are of course clean surfaces and liquids, no modification of the surface due to the liquids and a negligible gravitational effect on the liquid drop. *Young* derived a relation of the different interfacial energies γ_{ij} between the three phases vapor (V), liquid (L) and solid (S) (see Fig. 2.1). The angle Θ is the so called contact angle.

$$\gamma_{LV} \cdot \cos\Theta = \gamma_{SV} - \gamma_{SL} \tag{2.6}$$

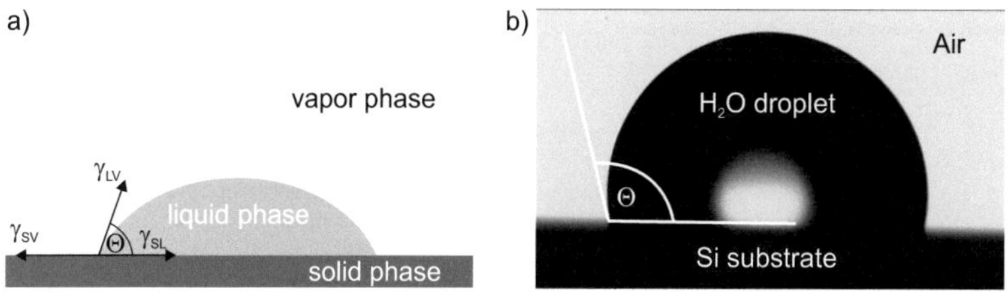

Figure 2.1: *a) Vectorial representation of surface tensions and energies according to Young and b) a water droplet on a hydrophobic silicon surface.*

The vapor phase consists not only of air but to a great extent of vaporized molecules from the liquid phase. The vaporized molecules may adsorb to the solid surface and reduce the effective SFE of the solid. Especially for high SFE solids, the amount of adsorbed vapor molecules cannot be neglected. The difference between the real SFE of the solid γ_S and the value γ_{SV} obtained from the Young equation is defined as spreading pressure π.

$$\pi = \gamma_S - \gamma_{SV} \tag{2.7}$$

Applying the Dupré equation Eq. (2.5) to the situation of a sessile drop on a surface and using the definitions in Eqs. (2.6),(2.7) gives

$$W_a = \gamma_{LV}(1 + \cos\Theta) + \pi. \tag{2.8}$$
$$(\text{with } \gamma_1 = \gamma_S,\ \gamma_2 = \gamma_L = \gamma_{LV})$$

Only in cases of low SFE-surfaces like fluorinated antisticking layers and high tension liquids, π can be neglected because the vaporization and adsorption of liquid phase molecules is small.

Eq. (2.9) allows a direct calculation of W_a from contact angle measurements and known liquid properties. If the interfacial tension γ_{SL} between the solid and the liquid phase as well as W_a would be known, a relation between the solid surface free energy and the liquid tension could be established by Eq. (2.5). This would in principle allow the calculation of the solid SFE from contact angle measurements and known liquid data. *Good* and *Girifalco* solved that problem by deriving an expression for the unknown interfacial tension γ_{SL}. First, they introduced the thermodynamic parameter Φ as relation of W_a between two materials 1 and 2 to the harmonic mean of W_c of both materials. Φ becomes unity for a homogeneous material without an interface.

$$\Phi = \frac{W_a}{\sqrt{W_{c1} \cdot W_{c2}}} \tag{2.9}$$

With the Dupré equation Eq. (2.5) and Eq. (2.4) sone obtains

$$\Phi = \frac{\gamma_1 + \gamma_2 - \gamma_{12}}{2\sqrt{\gamma_1 \gamma_2}} \tag{2.10}$$

or

$$\gamma_{12} = \gamma_1 + \gamma_2 - 2\Phi\sqrt{\gamma_1\gamma_2}. \tag{2.11}$$

One problem with this new equation is now the unknown parameter Φ. Three different approaches are discussed how to deal with that parameter.

Fowkes' Dispersion Approach This approach is based on the two assumptions made by *Fowkes* that i) the surface energy can be divided into different components stemming from the different intermolecular forces discussed in Sec. 2.1 and ii) that an apolar liquid can only interact with the apolar, i.e., the dispersion component of the solid in contact with the liquid. *Fowkes'* assumption led to todays *Fractional Polarity Theory* expressing the overall SFE as sum of the single contributions (d...dispersion, p...polar dipole-dipole, i...polar dipole-induced dipole, h...hydrogen bonding) (cf. [26, p. 140]).

$$\gamma = \gamma^d + \gamma^p + \gamma^i + \gamma^h \tag{2.12}$$

As stated by *Pocius* [23], this is only an approximation because potential energies cannot be summed up without accounting for cross-relations, which is not done in this case.

As discussed more detailed in Appendix A, one obtains for the SFE of low energy solids and high tension, dispersive liquid probes

$$W_a = \gamma_{LV}(1 + \cos\Theta) = 2\sqrt{\gamma_{LV}^d \gamma_{SV}^d}, \tag{2.13}$$

which allow the direct calculation of the unknown value γ_{SV}^d by rearranging the equation to

$$\gamma_{SV}^d = \frac{\gamma_{LV}^d}{4}(1 + \cos\Theta)^2. \tag{2.14}$$

To determine the polar contribution to the solids SFE, a liquid probe having a polar contribution as well is required. This extension of *Fowkes'* approach for a polar contribution is discussed in the following.

Owens and Wendt Approach This is a direct extension of the original idea of *Fowkes*. *Owens* and *Wendt* assumed that

$$\gamma = \gamma^d + \gamma^p \tag{2.15}$$

with γ^d including the already known dispersion forces and γ^p including all polar forces. As seen from Eq. (2.13), W_a can be in a first approximation calculated from a geometric mean of the interacting forces. This can be extended to the polar contribution. In other words, the work of adhesion is composed of the geometric mean of the different contributing fractional polarity components. Based on this assumption one obtains

$$W_a = \gamma_{LV}(1 + \cos\Theta) = 2\sqrt{\gamma_{LV}^d \gamma_{SV}^d} + 2\sqrt{\gamma_{LV}^p \gamma_{SV}^p}. \tag{2.16}$$

By rearranging this equation, one obtains an expression of the type $y = mx + b$:

$$\underbrace{\frac{\gamma_{LV}(1+\cos\Theta)}{2\sqrt{\gamma_{LV}^{d}}}}_{y} = \underbrace{\sqrt{\gamma_{SV}^{p}}}_{m}\underbrace{\sqrt{\frac{\gamma_{LV}^{p}}{\gamma_{LV}^{d}}}}_{x} + \underbrace{\sqrt{\gamma_{SV}^{d}}}_{b} \tag{2.17}$$

Plotting the left hand side of this equation, the normalized work of adhesion, against the ratio of the known liquid probe properties allow the extraction of the solids properties by a graphical approach (see Fig. 2.2). This methods was also used in this work to obtain γ_{SV}^{d} and γ_{SV}^{p}. At least two liquids of known surface tension components are required to determine the respective surface energy components of the unkown surface. In principle, an equation systems with two unknowns and two equations must be solved. Using more than two known liquids gives an overdetermined equation system, but improves the reliability of the determined surface energy. The overdetermined equation system can be solved in a graphical approach by a linear regression. This approach was the standard method applied in this work.

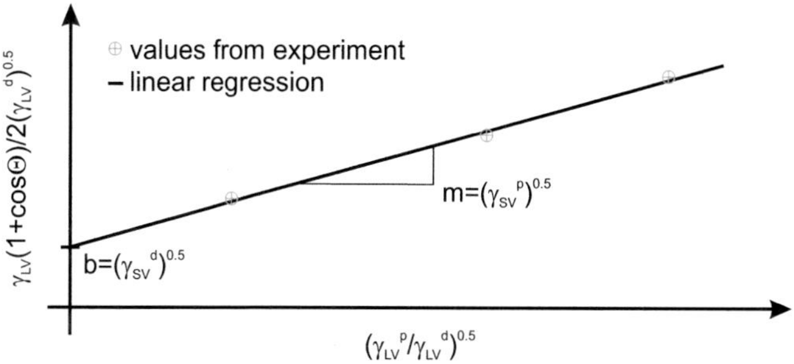

Figure 2.2: *Linear regression for the normalized work of adhesion for different probe liquids.*

Lifshitz-van-der-Waals Theory and Acid-Base Approach The following very short discussion is based on [26, pp. 137]. Details can be found there and only a brief overview of this theory is given. Consequences for NIL are discussed in Sec. 2.7.3. The fractional polarity theory as shown in Eq. (2.12) and the assumption of an independent additive interaction of the dispersion, dipole-dipole, dipole-induced dipole and hydrogen forces across material interfaces can be, exactly said, only used for gaseous phases. This additive approach is still valid for dispersion forces in condensed phases due to their isotropic, or scalar, character. W_a can be correctly calculated with Eq. (2.13) for dispersion forces in condensed phases. The vectorial character of dipole-dipole and dipole-induced dipole forces, however, means that these forces can partially cancel out each other. Thus, the resulting work of adhesion can be smaller than suggested by an independent, pairwise, additive combination for all fractional forces. A solution for this problem is provided by the *Lifshitz* theory combining dispersion forces with dipole-dipole and dipole-induced dipole forces as one force being responsible for a charge displacement in molecules. Basically, this theory treats the dipole-dipole and dipole-induced dipole as a second order effect and makes dispersion forces the prominent forces of the three mentioned forces. The newly defined Lifshitz-van der Waals (LW) contribution γ^{LW} to the surface free energy becomes

$$\gamma^{LW} = \gamma^{d} + \gamma^{p} + \gamma^{i}. \tag{2.18}$$

By using the Good-Girifalco interaction parameter [cf. Eq. (2.10)], the theoretical value of Φ^{LW} due to LW-interaction for two phases 1 and 2 is,

$$\Phi^{LW} = \frac{\gamma_1^{LW} + \gamma_2^{LW} - \gamma_{12}^{LW}}{2\sqrt{\gamma_1^{LW}\gamma_2^{LW}}}. \tag{2.19}$$

This value were found to be close to unity. However, experimentally obtained results for the interactions parameter were significantly different from unity when systems with known hydrogen bond interaction were present. The missing term in the surface free energy, accounting for an interaction parameter smaller than unity, was attributed to hydrogen bonding γ^{h}. Because hydrogen bonding is only one type of lewis-acid (A) and lewis-base (B) interaction the more general term γ^{AB} was used. The fact that the true interaction parameter is different from unity for materials with hydrogen bonds, or acid-base interaction, demonstrates that the bulk (or cohesive) behavior of the single materials are different from the interfacial (or adhesive) properties [cf. Eq. 2.9]s. It is today well accepted that acid-base interactions can significantly contribute to adhesion.

The group of *van Oss*, *Chaudhury* and *Good* (vOCG) introduced a set of parameters defining the acid-base properties of materials. Detailed discussions on the origin of these parameters can be found in [26] and are beyond the scope of this work. The vOCG-group postulated a combining rule according to

$$W_a = \gamma_{LV}(1 + \cos\theta) = 2\sqrt{\gamma_1^{LW}\gamma_2^{LW}} + 2\sqrt{\gamma_1^{+}\gamma_2^{-}} + 2\sqrt{\gamma_1^{-}\gamma_2^{+}} \tag{2.20}$$

which is mathematically equal to the additive interaction of the different forces but has by its theory an important difference: The acid-base components interact with its respective counter-component instead of the similar component. The parameter γ^{+} is related to the properties of a proton donor, in other words an electron acceptor or a Lewis-acid. The second parameter γ^{-} characterizes the proton acceptor, or electron donor, or Lewis-base properties of a material. An acidic component requires a basic component to interact. Thus, acid-base properties do not contribute to cohesion if the respective counter-component is not present in the bulk but might significantly contribute to adhesion if a material with the counter-characteristics is present. This explains the observed deviation of the interaction parameter from unity if hydrogen bonds, in other words acid-base interactions, are present on surfaces.

McCafferty [27] proposed a method similar to the graphical Owens-Wendt approach discussed above [cf. Eq. (2.17)]. He used two steps to solve the equation system for the three unknowns γ_{SV}^{LW}, γ_{SV}^{+} and γ_{SV}^{-} of the solid surface. Similar to the Fowkes principle, a test liquid with γ_{SV}^{+}=γ_{SV}^{-}=0 is used. This reduces Eq. (2.20) to

$$\gamma_{LV}(1+\cos\theta) = 2\sqrt{\gamma_1^{LW}\gamma_2^{LW}} \tag{2.21}$$

which allows for the direct calculation of the of γ_{SV}^{LW} like in the Fowkes approach [cf. Eq. (2.13))] from known liquid probe parameters.

$$\gamma_{SV}^{LW} = \frac{\gamma_{LV}^{LW}}{4}(1+\cos\Theta)^2. \tag{2.22}$$

Only two unknowns are left in Eq. (2.20). Two additional liquids with known acid-base parameters are required to solve the resulting equation system. An advantageous set of liquids should be chosen to solve this equation system [27]. To include results of more than two additional liquids in a "best-fit" approach and to solve the overdetermined equation system, again a graphical solution can be used. By rearranging Eq. (2.20) one obtains

$$\underbrace{\frac{\gamma_{LV}(1+\cos\theta) - 2\sqrt{\gamma_{LV}^{LW}\gamma_{SV}^{LW}}}{2\sqrt{\gamma_{LV}^{+}}}}_{y} = \underbrace{\sqrt{\gamma_{SV}^{+}}}_{m}\underbrace{\sqrt{\frac{\gamma_{LV}^{-}}{\gamma_{LV}^{+}}}}_{x} + \underbrace{\sqrt{\gamma_{SV}^{-}}}_{b} \tag{2.23}$$

or

$$\underbrace{\frac{\gamma_{LV}(1+\cos\theta) - 2\sqrt{\gamma_{LV}^{LW}\gamma_{SV}^{LW}}}{2\sqrt{\gamma_{LV}^{-}}}}_{y} = \underbrace{\sqrt{\gamma_{SV}^{-}}}_{m}\underbrace{\sqrt{\frac{\gamma_{LV}^{+}}{\gamma_{LV}^{-}}}}_{x} + \underbrace{\sqrt{\gamma_{SV}^{+}}}_{b}. \tag{2.24}$$

Both equations represent a linear equation in the form $y = mx+b$. The y- and the x-value is calculated for each test liquid from measured or tabulated values. A linear regression through these datapoints gives the unknown values as the slope and the intersection variable of the linear equation. Whether Eq. (2.23) or (2.24) is favorable depends on the available liquids and the tabulated values of the tension parameters. To improve the regression quality, the datapoints should not be bunched together and be rather equally distributed over the abscissa. This graphical solution also allows a direct visual comparison of different surfaces due to different curves in the linear plots.

One important aspect should be noted. All these approaches give only an approximation of the SFE mainly due to the fact that the spreading pressure is neglected and that surface tension parameters of test liquids are not very well standardized and differ between different groups. This should be always kept in mind when SFE values are compared.

2.4 Adhesion Models

Several models of adhesion are relevant for NIL. Basically, the imprint material can be seen as an adhesive, desired to adhere well to the substrate but not to the mold. The following models are based on overviews given by [22, 23].

2.4.1 Electrostatic Theory

Although electrostatic forces are known to play a role in adhesion, at least for surfaces with large differences in electronegativity [23], they are expected to play a minor role in nanoimprinting. Normally, no charged surfaces are present neither for the substrate nor for the imprint mold.

2.4.2 Mechanical Theory

Surface Roughness Surface patterns contribute in several ways to adhesion. The first is, that a microscopic surface roughness increases the effective area in contact with the imprint material for both the substrate and the mold. It is obvious, that the above discussed forces are related to the respective amount of molecules interacting with each other and, thus, scale with the area in contact. Microscopic roughness can lead to a tremendous increase in surface area and adhesion.

Mechanical Interlocking This occurs when the imprint material penetrates in pores or structures having an undercut geometry (see Fig. 2.3a). To separate the imprint material

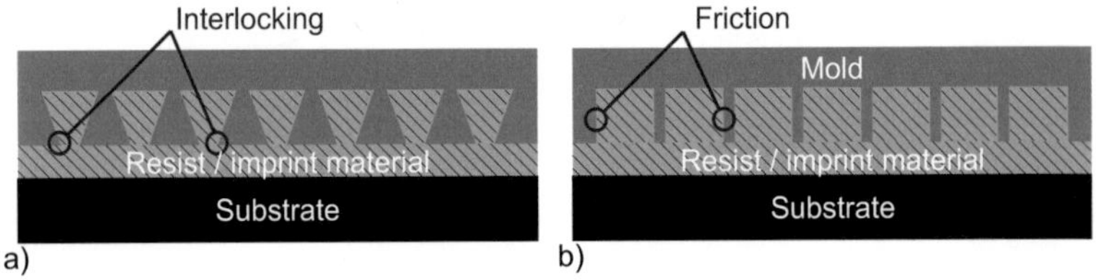

Figure 2.3: *a) Undesired NIL situation with undercut structures leading to interlocking and b) typical situation during NIL with friction on mold sidewalls during demolding (modified from [24]).*

from such an undercut requires a deformation or rupture of the imprint material, which dissipates a lot of energy resulting in an effectively increased adhesion. For nanoimprinting, smooth substrates with values below 1 nm RMS roughness are used. The RMS micro-roughness (sidewall roughness etc.) of the mold is determined by the mold manufacturing steps and should be also low for an optimal performance. Undercuts of etched features result from the respective etch recipe and should be avoided to prevent a rupture of the imprint material during demolding. Friction (see Fig. 2.3b) between the imprint material and the mold can only be avoided by resist shrinkage. Especially for high aspect ratios friction becomes a major failure reason.

Often discussed is the surface increase due to mold patterning. This can be illustrated by considering a checkerboard pattern (see Fig. 2.4). The overall contact area A is the sum of the area A_0 of the unpatterned mold and the additional sidewall area A_S of the etched grooves.

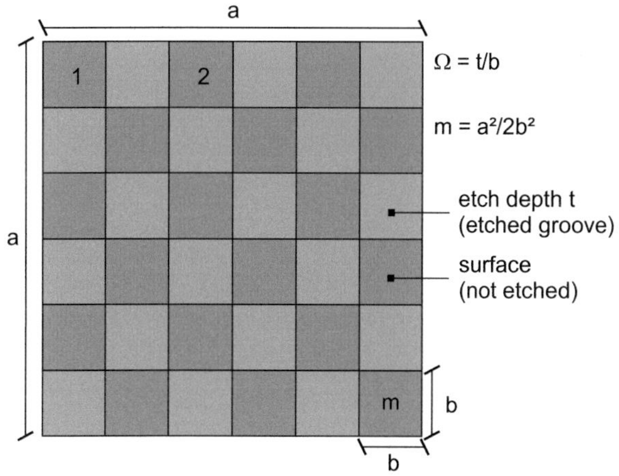

Figure 2.4: *Theoretical checkerboard pattern illustrating the relative increase of the surface area due to a patterned surface.*

$$A = A_0 + A_S = a^2 + 4btm = a^2 + 4b^2\Omega m = a^2 + 4b^2\Omega\frac{a^2}{2b^2}$$
$$A = a^2(1 + 2\Omega) \tag{2.25}$$

An aspect ratio of one gives already an overall contact area, which is three times larger than the unpatterned area. The aspect ratio does normally not exceed ten for typical applications. However, this leads to a huge increase of the overall contact area.

2.4.3 Diffusion Theory

The diffusion theory is based on the formation of an interphase instead of an interface between two materials, where at least one of the materials can diffuse into the other. Normally, the imprint material is not able to form such interphases with the mold or the substrate. However, special coupling agents or adhesion promoters can be applied to nanoimprint substrates forming a "soft" layer. It is thought that a partial diffusion of the imprint material into this layer is responsible for the improved adhesion observed for these coupling agents [23]. As diffusion is enhanced at higher temperatures, an annealing step can improve the adhesion on a coupling agent treated substrate.

2.4.4 Adsorption Theory

It is agreed on, that the adsorption theory cannot explain all effects of adhesion. However, for NIL it is the most obvious explanation for the observed adhesion effects. The basis for physisorption are the physical forces discussed in Sec. 2.1. Among them, dispersion forces are thought to be the most prominent forces. If polar groups or molecules are present in the adhesive, i.e. the imprint material, an orientation of the molecules on the interface are thought to play an important role. Such orientation processes in polymers are well known from the science of molecularly imprinted polymers [28]. Non-crosslinked, low molecular weight components have a higher mobility than long entangled or cross-linked polymer chains and, as a consequence, can more easily align and orient oneself on interfaces. This could be one reason for the observed effect, that in thermal based NIL (T-NIL) sticking is less pronounced than in UV-NIL [24]. However, alignment seems to occur more frequently on high energy surfaces. Physisorption explains the good wetting of surfaces being a prerequisite for good adhesion, even though a clear mathematical relation between wetting and adhesion is difficult to establish.

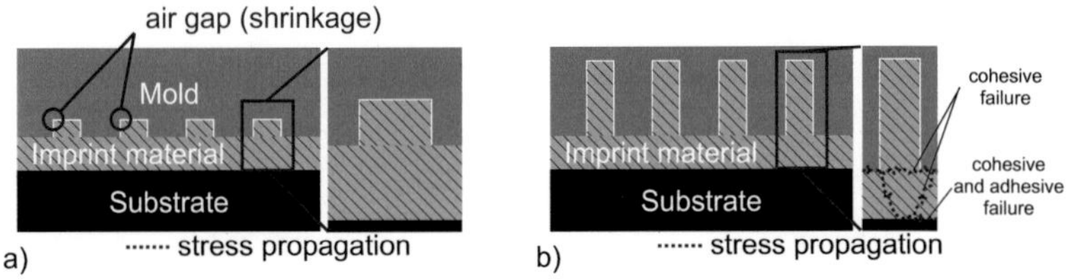

Figure 2.5: *Most likely routes of stress propagation a) for low aspect ratio features (= clean demolding) with a weak boundary layer due to imprint material shrinkage and b) for high aspect ratio features (= demolding failure).*

Air inclusions or unfilled cavities can lead to surface flaws, which act as stress concentrators and lead ultimately to adhesion failures. Such effects are known as weak boundary layer effects. In NIL, this is an advantage for the imprint material-mold interface. A separation along this interface is required for a clean mold release. In this context, a shrinkage of the imprint material enhances clean demolding by providing a "natural" line for stress propagation being favored over a propagation in the imprint material. However, if the aspect ratio of the mold becomes too high, a cohesive failure in the imprint material or an adhesive failure along the substrate can be the consequence because a "detour" of the stress along the interface with the mold is not favored any longer above the direct "tour" through the imprint material or along the substrate. Thus, a good wetting on the substrate and a poor wetting of the mold can be used as a rule of thumb for good imprint results. In this terms, wettability and SFE can be used to predict adhesion rather than explaining it.

2.4.5 Covalent Bonding

Even though physisorption can explain most of the observed strong adhesion effects, chemisorption is believed to be the true reason for such strong adhesion. Chemisorption results from the formation of covalent bonding, which is often achieved by special coupling agents (cf. Sec. 2.6). However, covalent bonding can also be achieved for suitable surfaces without coupling agents.

As mentioned above (cf. Sec. 2.3), acidic-surfaces can interact well with basic-surfaces. This was discussed in terms of adhesion by several authors [23, 26, 29]. The following tables give an overview on different groups and material surfaces with their acid-base properties (see Tabs. 2.2,2.3).

Table 2.2: *Lewis acid-base character of different groups explaining the acid-base character of surfaces discussed in Table 2.3*

Group	Exemplary surface	Ref.
	Bases	
Ester	PMMA	a
Ether	PEG	a
Hydroxyl (Oxygen)	PEG	a
Chlorine	PVC	a
Aromatic rings	PS	a
Secondary amine (-NH_2)	-	b
Primary amine (-NH)	-	b
	Acids	
Hydroxyl (Protons)	PEG	a

a: [26], b: [29, p.49]

The isoelectric point (IEP) of a metal oxide surface is a valid method to determine its acid-base character [29, p.48]. *McCafferty* compared the IEP of metal oxide surfaces and their acid-base properties with the measured adhesion strength. As a conclusion he could confirm that a basic polymer like polymethylmethacrylat (PMMA) adheres better on an acidic surface like silica than on a more basic surface like alumina. Vice versa the adhesion of an acidic polymer is better on basic surfaces [27, 30].

The determination of the acid-base character due to the γ^+- and γ^--values is more difficult than using the IEP. The reason is Lewis neutralization [26]. Protons of a Lewis-acid can react with Lewis-bases within the same material: This neutralizes the acidic character towards the outside of the material. For example, hydrogen atoms of hydroxyl groups can be neutralized by oxygen atoms of other hydroxyl groups or by other basic groups. Some examples should explain the problem (cf. Tab 2.3). PMMA is clearly a Lewis-base due to the ester groups. Polyethyleneglycol (PEG) 6000 is also a Lewis-base due to the ether groups. But PEG600 has a few hydroxyl groups in the structure, but

Table 2.3: *Lewis acid-base character of different surfaces relevant for adhesion in NIL.*

Surface	Character	γ_S* $[mJ/m^2]$	γ_S^{LW} $[mJ/m^2]$	γ_S^{AB}** $[mJ/m^2]$	γ_S^+ $[mJ/m^2]$	γ_S^- $[mJ/m^2]$	Ref.
			Polymers				
PMMA	base	39-43	39-43	0	0	9.5-22.4	a
PMMA	base	48.91	46.5	2.41	0.08	18.1	d
PEG6000	base	45	45	0	0	66	a
PVC	acid	43.75	43	0.75	0.04	3.5	a
PVC	acid	43.13	40.2	2.93	0.42	5.1	d
PS	acid	42	42	0	0	1.1	a
PVDF	acid	28.14-32.74	24.39-32.70	0.04-3.75	0.00-0.30	10.6-11.77	b
PVF	acid	39.78-46.61	36.20-46.46	0.15-3.58	0.00-0.36	7.93-8.90	b
PTFE	apolar	15.51-17.80	15.37-17.80	0.00-0.12	0.00-0.04	0.13-0.17	b
ETFE	apolar	17.22-20.02	16.51-20.00	0.13-0.71	0.01-0.13	0.80-0.96	b
			Oxides				
Cr_2O_3	acid	59.62	45.8	13.82	0.86	55.5	c
Al_2O_3	base	57.42	46.7	10.72	0.5	57.5	c
Si (nat. Oxide)	acid	40.5	35.9	4.5	0.2	25.3	d
SiO_2	acid	44.6	40.6	3.9	0.1	37.1	d

calculated with $^*\gamma_S = \gamma_S^{LW} + \gamma_S^{AB}$, $^{**}\gamma_S^{AB} = 2\sqrt{\gamma_S^+\gamma_S^-}$,
a: [26, p.165], b: [31], c: [27], d: [32]

its expected acidic character therefrom is not visible due to Lewis neutralization and the dominant ether groups. More difficult is polyvinyl chloride (PVC), which should be acidic due to its structure. However, only a small γ^+-value is visible for PVC, probably due to Lewis neutralization. As a result of Lewis neutralization, the acidic sites can be buried and might not point outwards. However, if a stronger Lewis-base will be present at the surface, the acidic site might turn towards this interface. This means, even though the acidic character might be not very visible, in combination with another material than the test liquid used to determine the acid-base character, the polymer might behave different. Another issue mentioned in [31], is that the acid-base approach by vOCG was found to overestimate the basicity of surfaces even when they should posses a clear acidic character. Lewis neutralization might be one explanation. A good example is polyvinylidene fluoride (PVDF) and polyvinyl fluoride (PVF), which are clearly acids. However, with test liquids used in the above mentioned study the behaved principally as bases to the outside.

As a consequence, the calculation of W_a and the comparison of different surfaces based on the measured acid-base properties are difficult. Furthermore, it is challenging and not simple to predict adhesion in a quantitative, theoretical manner.

2.5 Importance of Adhesion for NIL

Most important for NIL is the prediction and the tuning of adhesion both to the substrate and the mold. As shown, it is difficult to establish a general theory to connect the thermodynamical work of adhesion, which can be calculated from surface energy data, to the really observed, practical work of adhesion. However, it is possible to qualitatively compare different materials based on the thermodynamical work of adhesion. The main difference between different molds with the same layout are only the surface energies. This justifies such a comparison.

In principle, the adhesion to the mold and the substrate has not to be as low and as high as possible, respectively. The adhesion to the substrate must be just higher than the one to the mold. As seen above, the adhesion to a surface can be reduced by reducing the possibility to establish bindings. This could be achieved either by reducing the overall SFE or by providing surfaces which have incompatible surface characteristics (e.g., acidic surface and acidic imprint material). The first is usually achieved by using perfluorinated surfaces. Such surfaces have SFE values below $25\,\mathrm{mJ/m^2}$ depending on the molecule structure (see Tabs. 2.3,2.4). Perfluorinated surfaces with CF_x-compounds are almost purely dispersive surfaces and have the lowest possible SFE. Surface with high amounts of -CF_3 are the ultimate goal for antisticking coatings when SFE is considered. This led to the current state of the art antisticking layers (see Sec. 2.7,2.8). On the other hand, e.g., highly basic metal oxide surfaces would be useful as antisticking layers for highly basic polymers which would only adhere well to highly acidic surfaces like SiO_2.

Table 2.4: *Surface free energies of important surfaces terminated by hydro- and fluorocarbons.*

Group	SFE γ_{SV} [$\mathrm{mJ/m^2}$] [33]	SFE γ_{SV} [$\mathrm{mJ/m^2}$] [34]
-CH_2	-	36
-CH_2-	31	-
-CH_3	22-24	30
-CF_2-	18	23
-CF_3 and -CF_2-	17	17
-CF_3	6	15

The adhesion to substrates is normally increased by using adhesion promoters or coupling agents providing the best suited surface for the used imprint material regarding SFE and available adsorption sites.

2.6 Adhesion Promotion

Coupling agents are reactive primers enhancing the adhesion of two surfaces to each other. In NIL, these surfaces are the inorganic substrate and the organic imprint material. The task of primers is to tune the substrate surface to be beneficial for adhesion.

Coupling agents achieve this goal by creating an interphase between substrate and imprint material. Therefor, the coupling agent has to be difunctional. One functionality allows the chemical adsorption to the substrate. The second functionality enhances the adsorption and bonding of the imprint material to the coupling agent. The latter one could be due to diffusion (cf. Sec. 2.4.3) or physisorption (van-der-Waals interaction) or due to chemisorption (bond formation). The most common coupling agents are silane-based systems having molecules with a tailgroup, spacer and headgroup (see Fig. 2.6). The tail tunes the interfacial properties towards the imprint resist. For coupling agents this is mostly a organofunctional group realizing the link to the imprint material. The head group realizes the link to the substrate. Further details are discussed in Sec. 2.8. In contrast to physical bonds, chemical bonds cannot be replaced by water without destroying the bond itself. Physical bonds are reversible and can be displaced by water. Especially high SFE materials are energetically favored to be covered with a thin water layer than by e.g. a polymeric layer. This results in a displacement of the polymer and a weak adhesion of the polymer to the high energy surface. Stability against moisture is the main rationale for coupling agents.[22, 23, 26]

The second rationale for using reactive coupling agents during the course of this work was the high bond strength due to the chemical bond to the imprint material. It should be noted, that this chemical link is more or less agreed on even though direct experimental evidences are missing [26, p. 302].

2.7 Adhesion Prevention

2.7.1 Silane Antisticking Layers

Silanes are well known as coupling agents. Similarly, perfluorinated silanes have become the fundament of adhesion prevention or antisticking coatings for NIL due to their low SFE. The most important advantage of silane layers are the good adhesion to silica surfaces due to chemical adsorption and a quasi-monolayer characteristic if applied with an adequate coating process. This allows a feature conformal coating without structure distortion even for high aspect ratio features. Chemical adsorption also enables quite stable and robust coatings. A transfer of fluorine to the imprint material was only observed for silane molecules weakly bound (no chemisorption) to the molds surface and only during the first imprints (cf. Sec. 2.9.3).

2.7.2 Polymeric Antisticking Measures

Another way to provide fluorinated surfaces are polymeric coatings. Such coatings can be applied by spin-coating of solvent-polymer solutions. Drawbacks are the layer thicknesses of the respective layers well above the silane layers and the non-conformal coating. Especially small structures are difficult to coat. The usually thick layers lead to a structure distortion. Due to the low SFE of these layers, additional processes are required to achieve a good adhesion of these antisticking materials to the mold. Fluorocarbon layers can be applied for example by plasma deposition [35], a process quite similar to the plasma passivation process of sidewalls during cyclic deep reactive-ion-etching (RIE) processes. Thin layers can be also prepared by sputtering [35]. These layers can be prepared quite thin but still lack a sufficient adhesion to the mold leading to a fast antisticking-layer degradation and an imprint structure contamination.

More promising polymeric anti-sticking measures are polymeric molds which have intrinsic bulk anti-sticking properties [20, 36–44]. This intrinsic antisticking behavior means that no structural distortion due to additional layers occur and that no degradation of superficial layers limit the usability of the mold. An important issue for soft molds are structure deformations due to their reduced elastic modulus especially for high aspect ratio features and high imprint forces [45]. Another issue is the diffusion of several organic solvents and monomers into polymeric molds [46] which was also observed during this work. Perfluoropolyether derivatives (PFPEs) is a promising alternative to the often used polydimethylsiloxane (PDMS) due to the higher inertness against solvents [38] and resist constituents. The Young's modulus of PFPE is also higher than that of standard-PDMS. However, the PDMS-modulus can be also tuned by synthesizing PDMS with denser crosslinks [47].

2.7.3 Metalic Antisticking Layers

Finally, also metal oxides can be used as antisticking layers as discussed in the acid-base theory above. In T-NIL, metal coatings or metal molds are used due to the fact that polymers like PMMA adhere only weakly to surfaces like Cr, Al and Ni [24, 48]. This can be explained by the Lewis-base character of PMMA and the increasing Lewis-basicity of the superficial oxides of Cr, Al and Ni. Ni seems to be most effective [24] reflecting the fact that it forms the most basic oxide of the mentioned metals. Also an AuPd-alloy was reported as anti-sticking layer [49] reducing interfacial bonds due to its noble character and, thus, low reactivity. As will be discussed later (see Sec. 2.9.3), low reactivity of the antisticking layers is important for their long-term stability.

Assuming the desired weak adhesion to the imprint material, transfer of metallic parts from the anti-sticking layer to the imprint material like for polymeric coatings are not expected. The adhesion of the oxide to the metal and the mold should be higher than the adhesion to the imprint material. The presence of weakly bound constituents like for polymeric coatings or certain silane coatings are not expected for metal oxide surfaces. The transfer of parts of the metals oxide to the imprint material is quite improbable.

2.7.4 Other Antisticking Measures

Other low-reactive surfaces due diamond-like [50] and fluorinated diamond-like [51] coatings were also used as alternative antisticking layers.

2.8 Silane Monolayer Preparation

Silanes are characterized by the mentioned head, spacer and tail groups (cf. Fig. 2.6). Whereas the head and tail influence the link to the substrate and the environment, the spacer is responsible for the alignment and the structure of the layer [52].

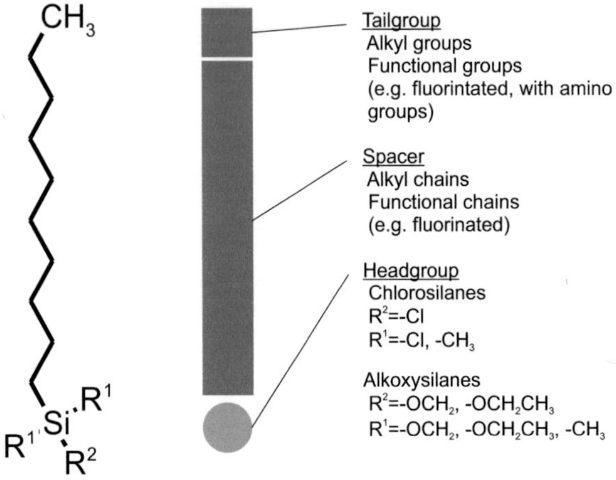

Figure 2.6: *Basic structure and principle representation of a decyltrichlorosilane molecule with a simple methyl functionality as tailgroup.*

Silanes have a central Si atom bonded at least to one hydrolyzable group which realizes the chemical siloxane bond to the substrate. The remaining Si-valences are usually bound to two more hydrolyzable groups of the same kind as the first one. This gives so called tri-silanes. These hydrolyzable groups can be, e.g., -Cl or -OCH_3 and -OCH_2CH_3 giving alkylchlorosilanes and alkylalkoxysilanes (methoxysilanes or ethoxysilanes), respectively. Chlorosilanes are known to be the most reactive ones in terms of hydrolyzation rates. They can even react directly with hydroxyl groups to form siloxane bonds, which is not possible with alkylsilanes [52]. However, high temperatures and chlorosilane concentrations seem to be required for that [53].

There are discussions about the structure of silane layers (cf. Sec. 2.8), but it seems to be agreed on that a typical silane reaction with a hydroxy-terminated surface like silica or metal oxides comprises three main steps (see Fig. 2.7). First, the reactive head group becomes hydrolyzed in the presence of water forming silanol groups (see Fig. 2.7a). This reaction can either happen in solution or at a water film at the substrate surface. Secondly, silanols can react with other silanols or hydroxy-terminated surfaces to from hydrogen bonds (see Fig. 2.7b and c). Third, once the hydrogen bond is established, a

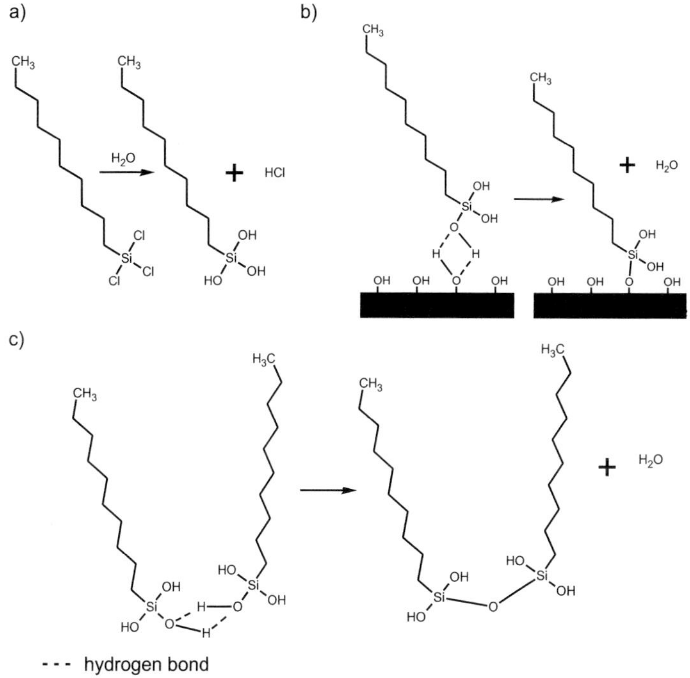

Figure 2.7: *Reaction of chlorosilane groups with a) water to a silanol via hydrolysis and subsequently with b) a hydroxylated surface or c) another silanol group to a siloxane bond.*

dehydration or condensation leads to the final linkage of the silane to the substrate by releasing water. This forms a siloxane bond in case of a silica substrate or an oxane bond on metal oxides.

Monolayers are beneficial for antisticking layers as they minimally modify feature sizes. Silane layers are quite common to tune surface properties (cf. Sec. 2.6). Typical antisticking layer silanes as 1H,1H,2H,2H-perfluorodecyltrichlorosilane (FDTS) or 1H,1H,2H,2H-perfluorooctyltrichlorosilane (FOTS) have fluorinated alkylchains and fluorinated tailgroups to reduce the SFE of the surface they are attached to. There are still discussions on the fundamental structure of these layers and if they can be seen as being truly self-assembling. *Stevens* discussed that trifunctional aklylsilanes can yield a full surface coverage but that such a coverage precludes a high degree of crosslinking

[54]. This is quite interesting because crosslinking is thought to be an important issue of silane layer stability [26, 54]. Currently, the silane layer seems to be imaginable as quasi-monolayer with many defects and some amount of crosslinks (see Fig. 2.8) [52].

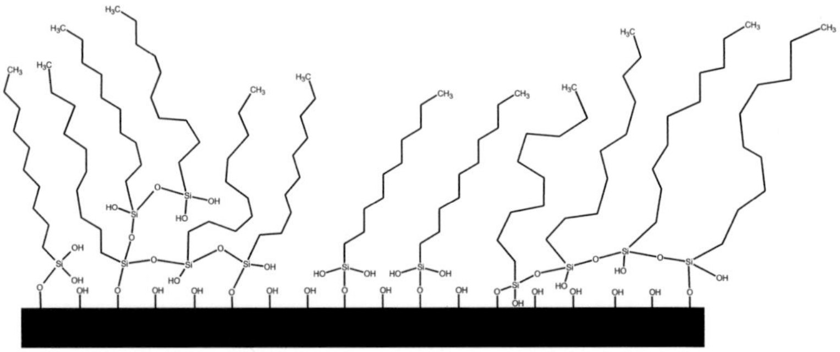

Figure 2.8: *Schematic representation of a silane quasi-monolayer including several defects and crosslinks between adjacent silane molecules (modified from [52]).*

2.8.1 Liquid Deposition

This technique is theoretically quite simple. The silane is dissolved in a solvent and the imprint mold is immersed in the coating solution. The quality of the monolayer depends mainly on the silane concentration, coating temperature, used solvent, substrate preparation and the immersion time [55, 56]. The high reactivity of silanes with water can already yield agglomerates in the coating solution. These agglomerates are deposited on the substrate and can even be covalently bound there due to silane crosslinking [56]. To avoid such agglomerates, a highly anhydrous environment must be provided, including dried labware, anhydrous solvents and freshly prepared silane-solutions [55]. Because chlorosilanes react with hydroxylated surfaces, glass or other labware, that posses also in a equilibrium state superficial hydroxyls, can already influence the effective silane concentration. Even though particulate free surfaces might be achievable with an optimized liquid process, the liquid deposition process is more difficult to control than a vapor phase process. Furthermore, the wet process leads to an inferior coating of deep and narrow nanostructures on the mold due to difficulties to wet these structures [56] and vapor coating is commonly favored for NIL.

Optool DSX is a commercial silane solution being often applied as liquid coating solution. Due to the much longer chain of Optool compared to FDTS or FOTS, the silanes are expected to lie on the surface instead of having an upright position [57]. Thus, Optool layers are in general thicker than FDTS or FOTS layers, which can explain the sometimes mentioned lower SFE of Optool coated surfaces. Due to the named problems, Optool should be not chosen for nanostructures.

2.8.2 Vapor Phase Deposition

This technique requires the transfer of the silane to be coated from its liquid pre-cursor state into the gaseous phase. The gaseous phase is the main reason for the superb coating results of vapor phase techniques. Due to the low density in the vapor phase, silane molecules cannot agglomerate and react almost exclusively on the substrate [52, 56]. They also coat high aspect ratio features and small nanostructures better than liquid phase approaches [56]. Undercut structures and even the bottom of substrates are well coated due to the migration of the vaporized molecules into the smallest structures and the absence of slow liquid diffusion processes [53]. Vapor coating is self limiting. As soon as no accessible hydroxyl groups are present, no more silane molecules are added to the growing layer [53]. The silane can be used as received liquid without dissolving it in a solvent. This makes the vapor coating very efficient. Already agglomerated silanes cannot be vaporized and remain in the pre-cursor container [52]. This was confirmed during this thesis.

Vaporization of the typically used silanes FOTS and FDTS at moderate temperatures requires a reduced pressure (see Tab. 2.5).

Table 2.5: *Overview on boiling points (bp) of common silanes required for vapor phase deposition .*

Short name	bp [°C]	p_{bp} [mbar]	Ref.	bp_{30mbar} [°C] ***
Propyltrimethoxysilane (PTMS)	142	1013.25*	a	-
Propyltrichlorosilane (PTCS)	123	5.33	a	-
10-undecenyltrichlorosilane (UDTCS)	100	1.07	a	-
3-(Trichlorosilyl)propyl methacrylate (TCSP-MA)	96-98	1013	b	-
Perfluoroctyltrichlorosilane (FOTS)	192	1013**	b	90
	84	22.66	c	-
Perfluorodecyltrichlorosilane (FDTS)	216-224	1013.25	a, b, c, e	141-144
	70	0.27	d	-
Octenyltrichlorosilane (OcTCS)	223-224	1013	b	-
Octadecyltrichlorosilane (ODTCS)	160	4	b	-

* atmospheric pressure, ** assumed, data missing,

*** with Clausius-Clapeyron relation: $\ln\left(\frac{p_2}{p_1}\right) = \frac{\Delta H_{vap}}{R} \cdot \frac{T_2 - T_1}{T_1 \cdot T_2}$

a:[52], b: ALDRICH Datasheet, c: ABCR Datasheet, d: [58], e: FLUOROCHEM Datasheet

2.9 Characterization of Monolayers

2.9.1 Surface Free Energy

The quality of a fluorinated ASLs is commonly estimated by its surface free energy determined with contact angle measurements. ASLs with a high amount of fluorinated groups have a SFE well below 20 mJ/m^2 (cf. Tab. 2.4). The SFE value cannot determine whether a monolayer or a multilayer is present but it can give some hints. A monolayer might not be perfectly dense and a contribution of the underlaying substrate to the SFE might be present. A multilayer system will have no contribution of the underlaying substrate. Thus, SFEs of multilayers will be somewhat lower than that of monolayers. This can be easily proven by comparing a vapor phase and a liquid phase deposited ASL with a high silane concentration or coating time during the liquid process. The vapor process is self-limiting whereas quite thick, white layers during the liquid phase can be achieved. The SFE of the liquid phase will be lower than that of the vapor phase.

The SFE indicates only the coverage of the surface with silane or other surface active components. *Cassie*'s law allows to determine the effective contact angle Θ_e of a heterogeneous surface with a material A and B having each a surface fraction of f_A and f_B and a contact angle of Θ_A and Θ_B, respectively.

$$\cos\Theta_e = f_A\cos\Theta_A + f_B\cos\Theta_B \tag{2.26}$$

Depending on the individual contribution of -CF_2, -CF_3 groups and a specific surface area coverage a certain SFE value will be obtained. The SFE cannot indicate whether the silane molecules are stably bound to a surface or not. This can be only determined by stability measurements.

2.9.2 Photoelectron Spectroscopy

X-ray photoelectron spectroscopy (XPS) is based on the photoelectric emission of so called photo-electrons due to the irradiation with X-ray photons. An incident X-ray photon transfers its energy to an electron which can leave the irradiated atom. The electron interacts with the neighboring atoms and leaves the material surface if the binding energy and the work function is overcome. The measurement is conducted in ultra high vacuum and the electrons can freely move in the vacuum from the sample surface to the detector. The detector consists often of two hemispheres and an electrostatic system. The electrostatic system and the hemispheres are adjusted in such a way that only electrons of a certain energy can pass through and reach the secondary electron multiplier. By setting this pass energy, the complete energetic spectrum can be scanned. The passing electrons are counted with a photomultiplier. The energy of the incident photon E_{Ph} is transferred to an electron with a certain kinetic energy E_{kin}. The kinetic energy depends on the binding energy E_B required to release the electron from the atom. Φ_S is the spectrometer work function, in other words the energy required to leave the surface; depending on the spectrometer and the material.

$$E_B = E_{Ph} - E_{kin} - \Phi_S \tag{2.27}$$

The photon energy for the used X-ray source in this work (ESCA 5700, Physical Electronics; Al Kα, not monochromatic) was E_{Ph}=1486.6 eV. Φ_S is known from calibration measurements. E_{kin} is determined by the above described detector by setting the pass energy and is the measuring variable of the setup. The binding energy is E_B characteristic for each element. Additional spectral signals arise from Auger-recombination, charging of the sample surface by electron depleted surfaces and X-ray satellites due to non-monochromatic X-ray sources. The energy of Auger-recombination is specific and known for all elements. Charging and the subsequent shift of the measured spectrum can be compensated using well known binding energies, e.g., that of the omnipresent carbon. X-ray satellites are also known and their effects can be removed from the spectrum. The atomic composition of a sample can be determined with a complete spectral scan. With a detailed scan of a certain binding energy region, the chemical compound composition and the bindings of a certain element can be determined. A chemical binding, e.g., that of silicon to oxide in SiO_2 shifts the binding energy of both elements from the value of a pure elemental composition to a new value of the compound. The binding energy a compound is also known and allows to determine the chemical compound composition of a sample. It is not always explicitly possible to discriminate the compounds as several compounds might overlap giving ambiguous results. [59]

A simple idea motivates angle-resolved XPS (AR-XPS) measurements. Whether an electron will be able to leave a surface or not depends on the line-of-sight distance (electron pathlength) between its origin and the sample surface. For the same origin in a given sample depth, this distance depends on the detector angle θ (see Fig. 2.9). θ can set by tilting the substrate.

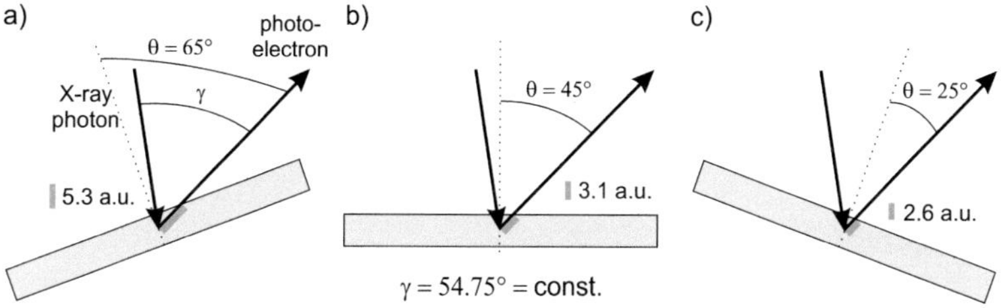

Figure 2.9: *Different detector angles θ a)-c) result in a different electron pathlength from the origin to the surface for the same origin for all detector angles.*

The electron pathlength, and thus the sampling depth, is limited due to scattering in the substrate and is normally less than 10 nm. For large angles θ, the sampling depth is smaller than for small angles of θ. This means, large angles measure relatively more electrons from the surface than from the bulk compared to small angles (see Fig. 2.10).

This makes AR-XPS a technique which is even more surface sensitive than XPS alone. Grafting a fluorinated silane to a sample surface will in theory result in an increasing relative amount of fluorine with increasing AR-XPS detector angles θ. A typical approach to characterize such a covering layer by AR-XPS was developed by *Paynter* [60] and further extended by *Kozlowska* [61, 62] for surface contaminations. Explicitly for silanes, this method was established in cooperation with *A. Kleiner* [63] and *B. Adolphi* at our institute (see Sec. 4.1.2 for details and results).

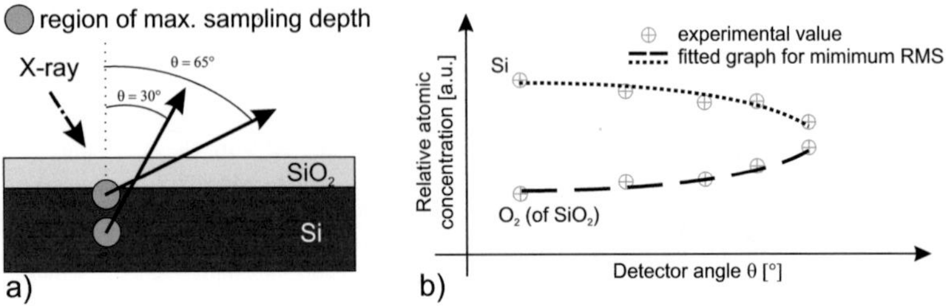

Figure 2.10: *Dependence of the maximum sampling depth on the detector angle θ: a) Larger angles promote a larger contribution of surface properties and b) resulting plot for relative atomic concentrations of the different elements of the material stack in a)*

The standard AR-XPS approach uses a parametric model, which is fitted to practical AR-XPS measurements of typical θ-values between 10°and 65°. The measurements give values as discrete points in Fig. 2.10b. The model gives a regression through all these points as output. Different parameters characterizing the surface layer can be extracted from the regression equation. A decomposition of elemental XPS peaks into single binding peaks improves the regression. The same is true for longer measurements times to improve the signal-to-noise ratio and the resolution of the single peaks. However, an influence of X-rays on fluorocarbon compounds is well known [57, 63, 64]. This requires a minimal measuring influence per detector angle to yield a constant sample surface constitution. The detector angle of the used ESCA 5700 system can be manually adjusted with an accuracy of ±0.5 °C.

Because AR-XPS is highly surface sensitive, the omnipresent hydrocarbon contamination of almost all surfaces is important. *Kozlowska* and *Kleiner* put many effort in modeling the surface contamination based on measurements. As the oxygen in the contamination is difficult to separate from the oxygen in the substrate, a clear composition of the contamination is almost impossible to obtain [62, 63]. Hydrochinone $C_6H_6O_2$ with a C:O-ratio of 3:1 is an example model for a surface contamination [62].

Kozlowska compared different modeling approaches [62] and chose the one by *Paynter* to be the most suitable. In the following, a short overview on the *Paynter*-model based on [61–63] is given. The model assumes that the surface is atomically smooth, the layers are homogeneous, the surface can be completely or partially covered by a surface layer expressed by the coverage F, photoelectron intensity is attenuated exponentially along the path from the origin to the surface, specific material attenuation coefficients do exist and neglects elastic electron scattering. The basic model considers an infinitely thick substrate being partly covered by a thin layer with the thickness d (see Fig. 2.11). It is crucial for the further understanding, that this cover layer has to be describable with a single chemical formula. This could be for example simply SiO_2, hydrochinone as mentioned above, FDTS or any other arbitrary composition reflecting the stoichiometry of the elements in the layer.

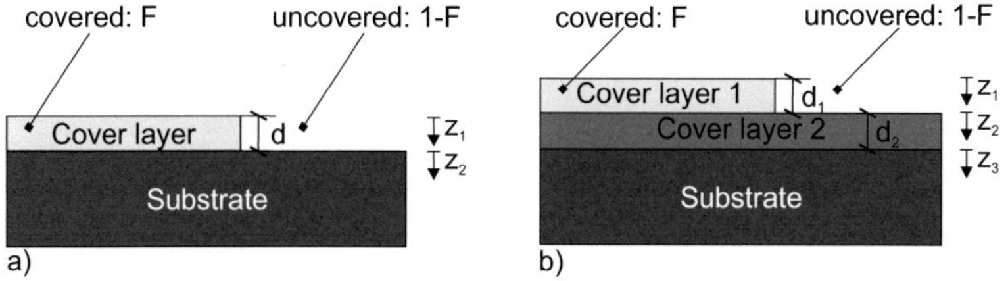

Figure 2.11: *a) Basic substrate incomplete coverlayer model with a layer thickness d and b) a three-layer model with a complete cover layer 2 and an incomplete coverlayer 1. Each model has a coverage fraction F.*

For a given chemical element x, the relation between the measured photo-electron intensity I_x and the sample specific parameters can be written as [60]:

$$\begin{aligned}\frac{I_x}{\phi T A \sigma_x} =& F\int_0^d c_{x,L}\exp\frac{-z_1}{\lambda_{x,L}\cos\theta}dz_1 \\ &+F\int_0^\infty c_{x,S}\exp\frac{-d}{\lambda_{x,L}\cos\theta}\exp\frac{-z_2}{\lambda_{x,S}\cos\theta}dz_2 \\ &+(1-F)\int_0^\infty c_{x,S}\exp\frac{-z_2}{\lambda_{x,S}\cos\theta}dz_2\end{aligned} \tag{2.28}$$

This is basically the sum of the different attenuated intensities of photoelectrons originating either

- from the cover-layer (covered region),
- from the substrate and passing through the cover layer (covered region)and
- directly from the substrate and passing not through the cover layer (uncovered region).

The left hand side of Eq. (2.28) is a normalization of the measured intensities to spectrometer specific constants: ϕ is the X-ray flux. T is the combined transmission/detector efficiency. A is the analysis area. σ_x is the cross-section for photoelectron emission. The right hand side of the equation are sample specific parameters. F is the surface coverage with the cover-layer[1]. $c_{x,L}$ and $c_{x,S}$ are the concentrations of the given chemical element in the cover-layer and in the substrate, respectively. $\lambda_{x,L}$ is the inelastic mean free path of electrons of element x in the cover-layer and $\lambda_{x,S}$ is the substrate analog.

[1]expressed as relative value of the covered area to the complete area

By evaluating the integrals in Eq. (2.28) for the given borders one obtains

$$\frac{I_x}{\phi T A \sigma_x \cos\theta} = F\left[c_{x,L}\lambda_{x,L}\left(1-\exp\frac{-d}{\lambda_{x,L}\cos\theta}\right)+c_{x,S}\lambda_{x,S}\exp\frac{-d}{\lambda_{x,L}\cos\theta}\right] + (1-F)c_{x,S}\lambda_{x,S} \tag{2.29}$$

1st Modification by Kozlowska According to *Kozlowska* [61], in Eqs. (2.28) and (2.29) derived by *Paynter*, the asymmetry factor L is missing in the denominator of the left hand side. The additional factor D in the *Kozlowska* formula-system is included in the factor T in the *Paynter*-equations. Two more important modifications of the *Paynter* approach were made by *Kozlowska*. First, the concentrations $c_{x,S}$ and $c_{x,L}$ from *Paynter* are substituted by the atomic densities $N_{x,S}$ and $N_{x,L}$ of the investigated element x in the substrate and the cover-layer defined as

$$N_{x,i} = \frac{C_{x,i}}{a_i^3} \quad \text{with}\, i = S, L. \tag{2.30}$$

$C_{x,i}$ is the XPS-measured atomic concentrations. This approach aims to consider different atomic densities in the sample structure [61]. a_i^3 is the average atomic volume of the respective layer i defined as [62]:

$$a_i^3 = \frac{M}{\rho n N_A} \tag{2.31}$$

M and ρ are the averaged molecular weight and the density of the basic building block of the cover-layer. n is the amount of atoms of this building block. N_A is the Avogadro constant. As said, a stoichiometrical description is required for the layer to define the basic building block. For SiO_2, n would be 3, M is 59.967 g/mol and ρ is 2.65 g/cm^3 [63]. Only $\lambda_{x,L}$ and $\lambda_{x,S}$ are left to be defined. They can be obtained from by using the attenuation length instead of the inelastic mean free path [62, 65]:

$$\lambda_{x,i} \approx 0.316 a_i^{3/2}\left[\frac{E_{kin}}{Z^{0.45}\left(\ln\frac{E_{kin}}{27}+3\right)}+4\right] \quad \text{with}\, i = S, L \tag{2.32}$$

E_{kin} could be obtained from Eq. (2.27).

2nd Modification by Kozlowska The second modification done by *Kozlowska* is the replacement of the normalized intensities [left hand side Eq. (2.29)]. With S_x as the standard XPS single-element sensitivity factor,

$$S_x = \phi T A \sigma L \lambda_x \tag{2.33}$$

the left hand side of Eq. (2.29) is simplified and one obtains

$$\frac{I_x}{S_x}\frac{\lambda_x}{\cos\theta} = F\left[\frac{C_{x,L}}{a_{x,L}^3}\lambda_{x,L}\left(1-\exp\frac{-d}{\lambda_{x,L}\cos\theta}\right)+\frac{C_{x,S}}{a_{x,S}^3}\lambda_{x,S}\exp\frac{-d}{\lambda_{x,L}\cos\theta}\right] + (1-F)\frac{C_{x,S}}{a_{x,S}^3}\lambda_{x,S} \tag{2.34}$$

It should be noted that there is some confusion in the literature on the left hand side of the above mentioned equation, probably originating from different definitions of the sensitivity factor S_x. In any case, the left hand side is an expression of an intensity normalized by specific spectrometer coefficients and is termed arbitrarily P for further use. Thereby, one obtains finally the *Kozlowska*-model used in this work:

$$\begin{aligned} P_x &= \frac{I_x}{S_x} \frac{\lambda_x}{\cos\theta} \\ &= \left\{ F \left[\frac{C_{x,L}}{a_{x,L}^3} \lambda_{x,L} \left(1 - \exp \frac{-d}{\lambda_{x,L} \cos\theta} \right) + \frac{C_{x,S}}{a_{x,S}^3} \lambda_{x,S} \exp \frac{-d}{\lambda_{x,L} \cos\theta} \right] \right. \\ &\left. + (1-F) \frac{C_{x,S}}{a_{x,S}^3} \lambda_{x,S} \right\} \end{aligned} \tag{2.35}$$

The calculated, fractional atomic concentration $AC_{x,calc}$ of an element x is always expressed relative to all other n measured elements and is calculated from

$$AC_{x,calc} = \frac{P_x}{\sum_{j=1}^{n} P_j} \quad \text{with } j = 1...x...n \tag{2.36}$$

$AC_{x,meas}$ is the value delivered typically by XPS-programs. Thus, the fractional atomic concentrations are the link between the experimentally obtained values and the theoretical model. The model values $AC_{x,calc}$ can be now compared for all angles and all elements with the corresponding measured values $AC_{x,meas}$. The difference between the model calculations and the measurements is expressed as root-mean-square (RMS) deviation and summed up for all n elements and all m angles:

$$\sum RMS = \frac{1}{m} \sum_{j=1}^{i} \sqrt{\frac{1}{n} \sum_{l=1}^{n} \left(AC_{x,calc|j,l} - AC_{x,meas|j,l} \right)^2} \tag{2.37}$$

An optimization algorithm can be used to minimize $\sum RMS$. The parameters F, $C_{x,L}$, $C_{x,S}$, d can be adjusted by the algorithm and are correlated to each other. Value constraints for the adjustable parameters should be used to obtain reasonable results. *Paynter* and *Kozlowska* used the commercial solver delivered with Microsoft Excel. The underlaying algorithm is the non-linear "Generalized Reduced Gradient (GRG2)" algorithm by *Lasdon* and *Waren* [60, 61].

The *Paynter* model was further extended by *Kozlowska* to a three-layer model with the layers $i = 1, 2, 3$, whereby only layer three has a partial coverage (cf Fig. 2.11b). Similar to the above derived equations one obtained:

$$\begin{aligned} P_x = \frac{I_x}{S_x}\frac{\lambda_x}{\cos\theta} = \Bigg[F \Bigg[& \frac{C_{x,1}}{a_1^3}\lambda_{x,1}\left(1-\exp\frac{-d_1}{\lambda_{x,1}\cos\theta}\right) \\ & + \frac{C_{x,2}}{a_2^3}\lambda_{x,2}\left(1-\exp\frac{-d_2}{\lambda_{x,2}\cos\theta}\right)\exp\frac{-d_1}{\lambda_1\cos\theta} \\ & + \frac{C_{x,3}}{a_3^3}\lambda_{x,3}\exp\frac{-d_1}{\lambda_{x,1}\cos\theta}\exp\frac{-d_2}{\lambda_{x,2}\cos\theta}\Bigg] \\ + (1-F) \Bigg[& \frac{C_{x,2}}{a_2^3}\lambda_{x,2}\left(1-\exp\frac{-d_2}{\lambda_{x,2}\cos\theta}\right) \\ & + \frac{C_{x,3}}{a_3^3}\lambda_{x,3}\exp\frac{-d_2}{\lambda_{x,2}\cos\theta}\Bigg]\Bigg] \end{aligned} \tag{2.38}$$

2.9.3 Stability of Silane Layers

When silanes are deployed as adhesion promoters or as an antisticking measure, a high stability against typical imprint process parameters and conditions is required.

Adhesion Promoters Reactive silane adhesion promoters like 3-(trichlorosilyl)propyl methacrylate (TCSP-MA) are very common adhesion promoters, especially for UV-curing resist systems. It is known, that a dehydration bake of the TCSP-MA coated surface is detrimental to the pendant methacrylate group and thus adhesion, which depends on this reactive head group [66].

Antisticking Layers ASL-stability is relevant for mold cleaning and imprinting. A self-cleaning effect of NIL-molds was demonstrated [19] and routinely observed during this work. At a certain degree of contamination, self-cleaning not sufficient and a more rigorous cleaning is required. An ASL which is stable against the cleaning procedures would be desirable to reduce ASL-recoating efforts. During imprinting, a highly stable ASL is mandatory to prevent a mold contamination and imprint defects.

Wet Chemical Cleaning Organosilanes on silicon oxides were reported to be relatively stable against hot solvents, oxidation and acids [67, 68] but not against bases [67]. However, minor degradation during acid treatment was visible in [68] due to slightly modified contact angles. This would in principle allow a mold cleaning by hot solvents. Short cleanings with suitable acids might be also possible.

Plasma Cleaning Resist residues can be removed by plasma processes. Fluorinated organosilane ASLs can be effectively removed with a plasma process having a chemical (oxidation) and a mechanical component (sputtering) [57]. Probably the sputtering effect is of major effect on the silanes.

UV-NIL The stability of fluorosilane-based ASLs during UV-NIL is an important issue and object of current research. It is commonly agreed on the fact that fluorosilane based ASLs continuously degrade during UV-NIL. The reasons are, however, still under discussion. ASL degradation was reported quite early [69–71] and nullified the idea of ever-lasting ASLs. Often the degradation was recognized based on a loss of fluorine from the ASL and an increase in SFE or adhesion energy [50]. The degradation seems to be due to chemical interactions of the ASL with the UV-NIL resist [50, 72, 73]. *Houle* and co-workers proposed fluorine abstraction to be one possible reason [50]. In accordance with other groups, they propose also that not in all cases a fluorine transfer to the resist takes place and that volatile fluorine-compounds as reaction products of the degradation are released to the ambient and are not detectable on the resist [72, 74]. Another suggestion was an acidolytic attack of the siloxane bond [50]. This is in contrast to the findings of *Wasserman* [67] and *Yoshino* [68], who reported a quite good acid resistance. A minor acidolytic contribution might be, however, present, as this is also visible in the results of *Yoshino*. Oxidation of the substrate and the silane layer was recently proposed [75] as degradation effect, being in contrast to the oxidation resistance reported by the above mentioned groups. Also a mechanical degradation can be present if weakly bound ASL-molecules are present on the surface, as typically observed for long chain ASL-molecules like the commercial Optool leading to thick ASLs [72, 75]. However, only weakly bound molecules are removed and at least a stable bound (mono)layer should be left. Almost no amounts of fluorine were transfered to the resist for thin, quasi-monolayers with almost no amounts of weakly bound molecules [72, 74].

As discussed in Sec. 2.6, coupling agents are most important in the presence of moisture. Traditionally, a high degree of crosslinking is thought to be a prerequisite for stable siloxane bonds making tri-silanes the most useful coupling agent in the presence of moisture. Monomeric silanes are thought to yield a higher degree of crosslinks than oligomeric silanes. It was argued that siloxane bonds are easily hydrolyzed. But they also reform quite easily as well. It was then further argued that it is quite improbable that all links of a highly crosslinked molecule hydrolyze at the same time. Consequently, the balance between hydrolysis and condensation is important. To achieve highly stable and resistant layers in this context means to complete all condensations reactions, to reduce the water at the condensation interface and to achieve a high degree of crosslinking. [26, pp.279].

Even though the presence of crosslinking in silane layers is still discussed, a quasi-monolayer with crosslinks seems to be present (cf. Sec. 2.8). This would require a shift of the hydrolyzation-condensation equilibrium towards hydrolysis to remove complete ASL-molecules by siloxane bond breaking and fluorine must be found on the imprint material. Thsi is often not the case and a fluorine abstraction seems to be most probable.

In summary, the degradation and resist-ASL interaction is highly material specific [69, 74]. A combination of acidolysis, oxidation and most importantly fluorine abstraction seems to be present. Mechanical degradation as discussed above seems to be of minor importance if crosslinked, quasi-monolayers are present. Not only low SFEs but ASLs being low reactive towards the used imprint material are required in NIL. SFE is no universal parameter for low reactivity and is, thus, no useful criterion or guarantee for a high quality and long-term stable ASL [74].

It is important to note, that mostly liquid phase deposited silanes were used so far for degradation characterization. However, for high quality imprints, vapor phase deposition is more advisable (cf. Sec. 2.7). Differences in layer properties are probable due to the quite different deposition processes. For example, siloxane bond attack contribution might be different. However, fluorine abstraction might be similar for both deposition techniques. See Sec. 4.1.2 for further discussion.

2.9.4 Antidegradation Measures

Different approaches tried to counteract the ASL-degradation and to ensure a clean demolding. Fluorine was incorporated into the resist formulation to minimize the ASL-resist interaction and to refresh the ASL [73, 76]. Pentafluoropropane was also proposed as a surfactant layer between resist and mold to reduce release forces [77]. Other concepts exploited mold materials with inherent antisticking properties like fluorinated polymer molds or ASLs being more inert against the imprint material (cf. Sec. 2.7.2 to 2.7.4). Some alternative ASLs are opaque and limited in their applicability for UV-NIL requiring alternative imprint techniques (see Sec. 4.4).

3 Basics of Lithography for Functional Polymer Waveguides

"Leave the atom alone."

(E. Y. Harburg)

UV-NIL offers unique advantages to directly pattern functional UV-curing materials with high aspect ratios up to ten and a sub-300 nm resolution at the same time. Polymer waveguides and photonics are an ideal field for such UV-NIL direct patterning approaches. UV-NIL offers a higher throughput than the sequential electron beam direct writing and laser direct writing with a comparable resolution. Hybrid UV-NIL techniques combined with photolithography compensate drawbacks of NIL for mixed-size patterns and avoid residual layers.

Regarding low loss polymer waveguides, silicon UV-NIL molds are favored above silica molds because of the lower sidewall roughness and better process control during reactive ion etching. Bonded nanoimprint molds are a cost-efficient mold alternative.

The incorporation of functional additives to UV-curing material formulations is a promising approach for chemical functional, polymeric surface reliefs and waveguides. Chemical functional waveguides are required for bioconjugation and specific biosensing with a polymeric microring resonator.

3.1 Planar Polymer Waveguides

The principle of optical waveguides is based on the confinement and transport of light in a high refractive index material surrounded by a low refractive index material. The highly refractive, light guiding material is termed core material. The surrounding is termed cladding material. The waveguide dimensions depend on the degree of light confinement, the wavelength of the used light and the refractive indices of the used materials. Typical dimensions for single mode polymer waveguides and a wavelength of 1550 nm are in the low µm-range for height and width. Integrated or planar optics are based on waveguides realized on or embedded in a substrate material as support. This is the main difference to fiber optics using bendable fiber structures requiring no support. Normally, a cladding is used in planar optics to have a good light confinement (see Fig. 3.1a to c). If the refractive indices of the upper and lower claddings are the same, the waveguide has a symmetric guiding characteristic meaning that the electric field intensity is symmetrically distributed across the waveguide. Different indices will result in an asymmetric field distribution meaning that the light is confined closer to the waveguide-core interface with the higher refractive cladding. For an application where the electric field has to probe the environment, e.g., a test liquid or an analyte, normally an asymmetric waveguide is used (see Fig. 3.1d).

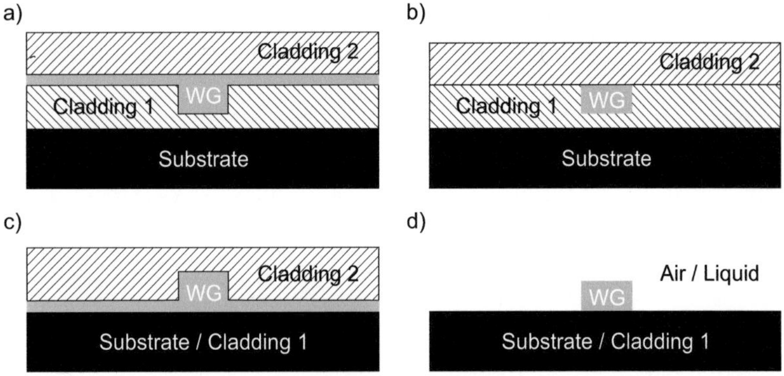

Figure 3.1: *Principles of planar waveguides with a) a ridge waveguide embedded into cladding materials and a substrate as a simple support, b) a technologically more challenging channel waveguide embedded into cladding materials, c) a ridge wavegide using the substrate as support and as cladding and d) a channel waveguide using the environment as cladding (WG .... waveguide).*

Different techniques can be used to produce such waveguides. The focus of the following sections is on the production of asymmetric ridge and channel waveguides without an additional upper cladding to be able to be in contact with a probe liquid (cf. Fig. 3.1d). The applicable techniques to realize such structures comprise i) mask-based and ii) mask-less or direct patterning techniques (see Fig. 3.2). A suitable technique depends on the required resolution, the structure dimensions, the waveguide material properties and the final application.

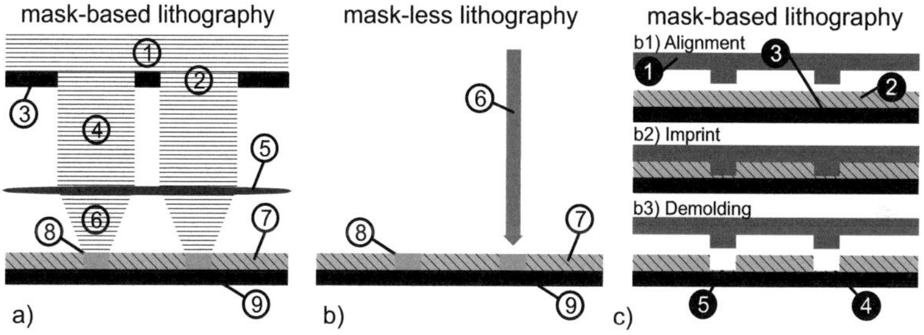

Figure 3.2: *Comparison of a) mask-based photolithography, b) direct-writing and c) nanoimprinting. (① flood exposure, ② transparent region of the photomask, ③ opaque region of the photomask, ④ transmitted exposure light, ⑤ optical lens, ⑥ convergent light, ⑦ photoresist, ⑧ photoresist with altered chemical properties due to exposure, ⑨ substrate; ❶ nanoimprint mold, ❷ nanoimprint material (resist), ❸ substrate, ❹ indented imprint material region, ❺ polymeric residual layer)*

3.2 Photolithography

For the system integration of optical devices like planar optics it is advantageous to use as much as possible of established semiconductor processes to stay compatible with electronics and benefit from grown knowledge. Photolithography is a well established and a matured technique for pattern definition.

In photomask-based lithography techniques, UV-light transfers the image from a photomask into a latent resist image. This latent image is composed of photoresist regions with different solubilities in a specific solvent. With these solvent the latent image can be converted into a real height topography by dissolving the more solvable regions. This step is also termed development.

The achievable resolution is limited by the particular technique being proximity, contact or projection lithography, the used wavelength λ, the used resist thickness d and the resist contrast. For proximity or contact lithography, the minimal resolvable structure width d_{min} [78] is

$$d_{min} = 1.5\sqrt{\lambda(G + d/2)} \tag{3.1}$$

with G being the proximity gap between photomask and resist. For contact lithography G equals zero. In projection lithography, the numerical aperture NA and a process dependent factor k determine the smallest resolvable structure width [79]:

$$d_{min} = k \cdot \frac{\lambda}{NA} \tag{3.2}$$

The photoresist topography has to be transferred in the optical waveguide material. This requires a polymer etching step, which implies challenges especially for functional polymers. Wet etching is not useful due to its isotropic character and its large amount of waste materials. Dry etching was demonstrated for polymeric, optical structures (e.g. [80]). Anisotropic, dry polymer etching requires a non-organic transfer material applied to the active optical material and an imaging material on top of the transfer material [81]. The imaging material is typically a photoresist. The etch selectivity between imaging and waveguide material is normally very low because both are organic materials. A transfer material is required to allow a highly selective etching. The pattern of the image material is transfered to the transfer material that has a sufficient etch selectivity against the waveguide material. The patterned transfer material is used as mask for etching the waveguide material. For this complex process, a material stack must be built on the optical material without disturbing its optical properties. O_2-plasmas with Ar support are required for highly anisotropic profiles [80, 81]. A result of this etch chemistry are often a redeposition of transfer mask material, polymeric material from the etch-bottom and substrate material due to overetching. Additionally, a crosslinking of the sidewalls can occur during transfer into the waveguide material [80, 81]. These effects are enhanced by the overetching time required in case of different structure densities and aspect ratios to clear all regions from the waveguide material except the waveguide itself. All these effects influence the functional groups of the waveguide material and are detrimental to functionalized waveguide materials that are based on superficial functional groups. A possible polymeric sidewall passivation is further difficult to be removed selectively from a polymeric waveguide. In this context, direct patterning approaches can offer advantages for patterning certain functional materials.

Direct Writing Direct writing is a serial, mask-less technique. Most drawbacks originate from the slow, serial processing. Energy sources like a focused laser or an electron beam are scanned over the substrate to directly write the latent image into resists or photo-definable materials. Direct writing can be used for both transfer masks and direct patterning of functional materials.

Electron Beam Direct Writing Electron beam lithography (EBL) is well established in microelectronics to pattern optical photomasks. EBL is based on the effect of electrons on a resist material, which could be "positive tone" (increase solubility in a developer solvent by chain scission) or "negative tone" (decrease solubility in a developer solvent due to crosslinking). EBL is normally used to define high resolution features in thin resist layers. The challenge of EBL for directly writing waveguides is the structure height. Most important is the scattering. Forward scattering occurs in all materials and leads to sloped waveguide sidewalls. Backscattering from the substrate influences features dimensions close to the substrate. To compensate these effects, higher acceleration voltages, proximity correction algorithms and high contrast electron beam resist can be used. [82]

For high acceleration voltages, expensive dedicated EBL-tools are required. High contrast resists interfere with the idea of direct writing into functional materials as the resist has to fulfill the requirements of the functional material and the EBL-resist at the same time, making the material quite complex. EBL uses limited writing fields below 1 μm^2. This requires a stitching of the writing fields for larger designs. For a high accuracy stitching also expensive tools and substrate stages are required.

EBL is a good prototyping technique and an important primary patterning technique for master structures like photomasks or nanoimprint molds. For the manufacturing of waveguides for a later system integration, a parallel mask-based technique offers a higher throughput and cost-reduction. NIL is one such possible technique and was chosen in this work.

Laser Direct Write Laser direct writing (LDW) uses the photo-reactions in the focal spot of a laser beam. The spot size can be smaller than the wavelength of the recording laser. The achievable resolution depends on non-linear optical effects, the resist contrast and the resists exposure threshold and can be even smaller than the focal spot size. After developing the exposed material, the topographical resist or waveguide pattern is obtained. The focal spot can be three-dimensional scanned through the resist and allows the patterning of arbitrary, three-dimensional features. [83]

3.3 Nanoimprinting

NIL is a dedicated polymer patterning technique. It is suited to directly shape functional device structures in polymers like MRRs. Two facts demand the use of nanoimprinting for such a polymer direct patterning: the resolution and the aspect ratio. For most functional device structures, high aspect ratios are required. For example, micron-sized polymer structures are required to guide light. However, evanescent light coupling requires feature distances in the sub-300nm region (cf. Fig. 1.2b). Such high aspect ratios, even in functional materials, are achieved with NIL in a "simple" molding or embossing step. NIL enables the patterning of sub-300 nm structures with high aspect ratios up to 10:1 [3].

Nanoimprinting is based on the displacement of an imprint material due to a mechanical force applied to the imprint material. To successfully transfer the mold pattern into the imprint material, it has to flow sufficiently under the used imprint parameters. The polymer flow is mainly determined by the viscosity of the imprint material under the used imprint conditions. Two basic NIL-techniques do exist depending on the chosen imprint material: hot embossing or thermal nanoimprint lithography (T-NIL) is used to pattern thermoplastic materials (see Fig. 3.3a). UV-assisted nanoimprint lithography (UV-NIL) is used for UV-curing polymers or UV-sensitive resists (see Fig. 3.3b). Instead of using UV-curing materials, also thermally curing materials can be used. Even though this is a thermally based technique, the material and the curing process itself resembles more a UV-NIL than a T-NIL process. There are also several combinations of these basic techniques possible. Good overviews on NIL-basics can be found in [24, 84–88]. In the following only a brief summary is given.

3.3.1 Thermal Nanoimprint Lithography (T-NIL)

As mentioned, NIL requires a viscous imprint material during the pattern replication step. In T-NIL, this viscous state is reached by heating thermoplastic materials above their glass transition temperature T_g. The flow behavior of a thermoplastic material depends highly on the temperature. At the microscopic scale, the thermoplastic material could be imagined as a network of linear, entangled polymer chains. Well below T_g, the material behaves almost elastic because the chains cannot be displaced permanently by the applied force. When the temperature increases, the polymer chains can move more freely. An applied force results first in an elongation of chain segments between entanglement points. This elongation is reversible and the material still behaves elastic. But to partially reconfigure entanglements and to enable polymer chains to move against each other, even higher temperatures than T_g are required. The respective temperature is termed terminal flow temperature T_f. This terminal flow region enables a permanent displacement of the polymer chains. The appropriate imprint temperature depends on the individual imprint material. The imprint material is usually spin-coated on the substrate (see Fig. 3.3a1). The mold and mostly the substrate as well is set to the imprint temperature and pressed into the imprint material (see Fig. 3.3a2). For stamp-and-repeat imprinting, the substrate temperature must be kept low enough to preserve the already imprinted features. After cooling to the appropriate demolding temperature, the replicated structure is released (see Fig. 3.3a3). Once the mold pattern is replicated, the temperature can be decreased and the pattern is "frozen" in the polymer. [24, 89]

T-NIL is less prone to resist sticking [24] and is thus in principle well-suited for high-aspect-ratio direct structuring. T-NIL also allows a post-imprint smoothening of polymer surfaces by tempering close to T_g [3]. This is useful for smooth, low-loss optical waveguides. Compared to UV-NIL, the polymer flow is more critical in T-NIL. Even at typical imprint temperatures well above glass transition it is still a highly viscous polymer flow with viscosities around 10^4-10^9 mPa·s [90, 91]. High pressures and temperatures are required to transfer the mold pattern into the polymer via T-NIL. This limits the applicability of T-NIL to certain heat resistant substrates, increases the process time and reduces the overlay accuracy of subsequent pattern layers if non-CTE-matching substrates are required. As a result, the optimization of the polymer flow and the residual layer could be challenging in T-NIL, especially for direct patterning using molds with small pattern densities, where large amounts of polymer have to be displaced.

3.3.2 UV-assisted Nanoimprintlithography (UV-NIL)

The low viscous imprint materials allows UV-NIL to be conducted at room temperature and much lower pressures than T-NIL. This increases the mold lifetime [92] and decreases the process time for the same mold-layout. The viscosity of the imprint material can be reduced by using low molecular weight materials like monomers or oligomers. For example, typical acrylate monomers can have viscosities down to 1 mPa·s. The imprint material can be spun-on or selectively dispensed (see Fig. 3.3b1). The mold is lowered into the liquid imprint material, which starts to fill the mold cavities. To preserve the mold pattern, the imprint material must be solidified. This could be achieved with curable moieties and a UV-flood-exposure (see Fig. 3.32). Mostly UV-curable formulations are used. Among them, free radical curing is most often used besides anionic or cationic initiator systems. After curing, the mold is separated from the imprint (see Fig. 3.3b3). UV-NIL requires a transparent mold or substrate to enable the UV-light to reach the imprint material. Mostly silicon substrates are used and thus a transparent mold is required.

The polymer flow in UV-NIL is less complex compared to T-NIL. However, depending on the mold layout, special measures against bubble trapping might be taken in account. Mixed pattern sizes are easier to replicate with UV-NIL due to the less complex polymer flow. Chemical functionalities can be easily incorporated in monomer formulations (cf. Secs. 3.4,4.2). Hence, UV-NIL is highly suited for customized polymer materials, which is often needed for functional microsystems. A further advantage of UV-NIL are the relative mild processing conditions. This is advantageous for functional materials to preserve the particular functionalities of chemical additives or other fillers.

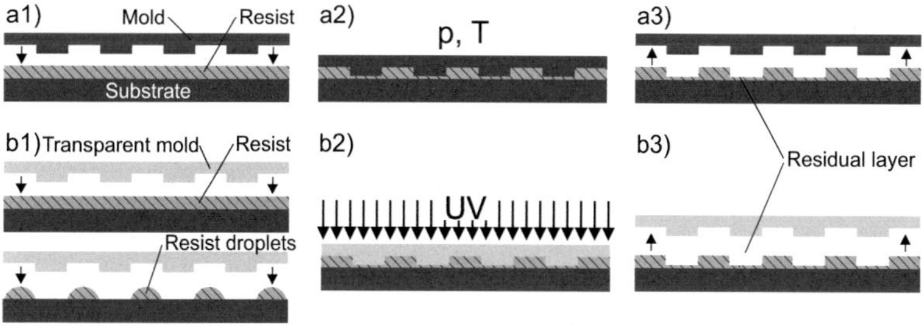

Figure 3.3: *Process flow for a) T-NIL and b) UV-NIL (a1: completely resist coated substrate, a2: pattern transfer at high pressures and temperatures, a3: cool-down and demolding; b1: completely coated substrate or dispensed pattern, b2: curing by UV flood exposure, b3: demolding).*

3.3.3 Hybrid Techniques

The combination of several complementary lithography techniques is a well known approach to balance drawbacks of individual techniques. For example, the fast but diffraction limited UV-photolithography (UVL) can be combined with ultra high resolution but comparable slow electron-beam lithography (EBL). In a similar way UVL can be combined with NIL. Individual advantages of hybrid techniques are

- simplified patterning of nm-scale and µm-scale features,
- avoiding the imprint-inherent residual layers and imprint post-processing,
- usability of the same resist for NIL and UVL (not possible for a UVL/EBL hybrid technique) and,
- increasing aspect ratios to a certain extent (residual layer control required).

T-NIL + UVL One major drawback of NIL refers to the simultaneous patterning of large and small features. Especially for high viscous imprint situations as T-NIL, small features will be imprinted much faster and easier than large features [24, 84]. This results in a mold deformation, an inhomogeneous residual layer distribution and some features failing to be resolveded (see Fig. 3.4) [93].

This feature size dependency can be resolved by splitting the patterning in two processes: one process for small, high resolution features and a second one for large features [48, 94, 95]. Many typical UV-photoresists exhibit a thermoplastic behavior [95]. This allows NIL and UVL to be performed in the same resist layer and avoids multi-resist processes as would be required for the combination of EBL and UVL. NIL is often used for the small features in the first step. This makes NIL-processing quite easy because no alignment to existing structures is required. The UVL alignment for the second step is mostly less critical. When the imprint is chronologically separated from the UVL step, a silicon T-NIL mold can be used. This implies less efforts during the mold manufacturing. Either positive or negative tone resists can be used and both require an optimized processing to minimize the interdependencies of the NIL and UVL steps [96]. Most challenging in this hybrid approach are the regions under the photomask edges. Typical effects are line end tailing and thinning, trenching and incisions [95]. Besides this, with thick initial and thus thick residual layers, even a multi-level patterning becomes possible [97].

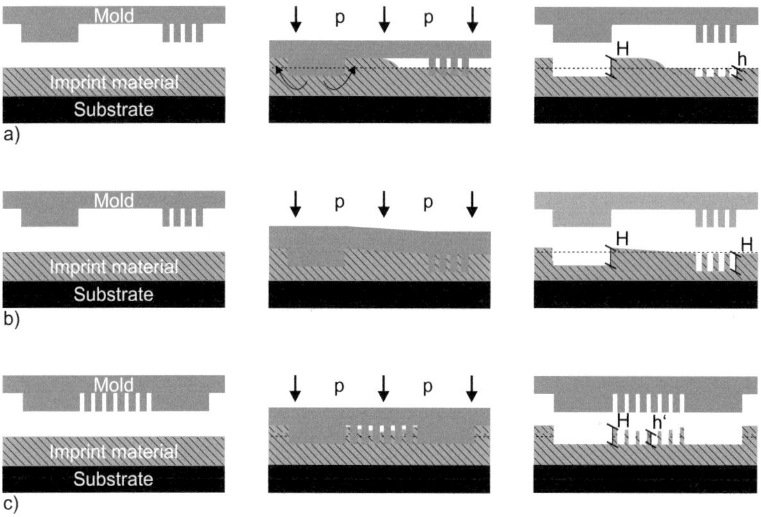

Figure 3.4: *Failures during NIL pattern replication with large and small features at the same mold leading to different feature heights H, h, and h': a) incomplete filling due to minimal material displacement by small features, b) mold deformation due to a faster imprinting of small features and a higher local effective pressure, c) incomplete filling due to the limited imprinting depth of large features. Note: The dashed line is the initial resist height. (modified from [93])*

UV-NIL + UVL It is also possible to combine the NIL- and the UVL-step in a single process step [93, 98, 99]. In such a single-step process mask edge effects as with the above mentioned two-step process can be avoided because NIL and UVL features are defined in the same step. The single-step technique is sometimes referred to as combined nanoimprint-and-photolithography technique (CNP). As a consequence of the single-step concept, hybrid nanoimprint molds (HNMs) are required. HNMs must be composed of UV-transparent and opaque regions, similar to a photomask. Additionally, HNMs must bear topographical features for the NIL-step. The particular HNM-design depends on the aim of the hybrid imprint technique. For example, HNMs dedicated for small and large feature patterning require large metal-regions and small topographical features (see Fig. 3.5a). HNM dedicated to avoid residual layers require a metal coating on the protrusions (see Fig. 3.5b).

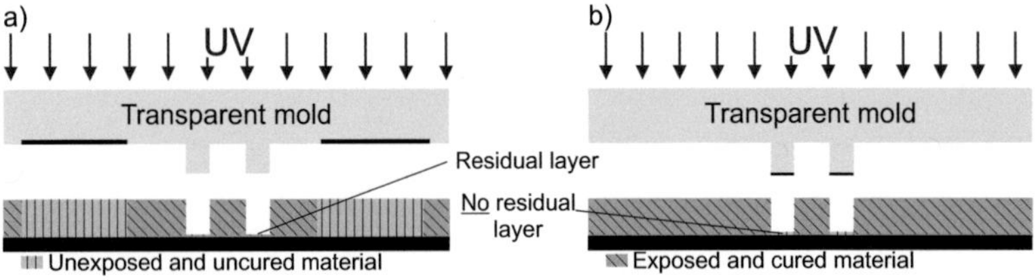

Figure 3.5: *Principle of CNP for the replication of a) large and small features at the same time and b) to avoid the residual layer under protrusions.*

The avoidance of residual layers is an important advantage of hybrid techniques. Under elevated mold features always a residual imprint material layer is present. This layer could be minimized but not completely squeezed away in finite times (cf. Eq. (3.3)) without using specific measures as modified surface free energies of the substrate, which normally interfere with imprint material adhesion to the substrate. In hybrid techniques, metal coatings on the imprint protrusions block the light from the residual layer [98, 100]. This technique can be used for UV-NIL materials or negative-tone photoresists. The unexposed residual layer can be removed easily by evaporating uncured and still low molecular weight materials, by a solvent rinse or a by developer process. When the residual layer could be well controlled, this technique enables also an increase in the aspect ratio as long as the residual layer is not exposed due to diffraction at sub-wavelength features (see Fig. 3.6). To successfully avoid an exposure of the residual layer under sub-wavelength features, its overall thickness has to be smaller than the exposure wavelength in the resist material [101]. An unexposed residual layer under sub-wavelength features could only be achieved in an optical near-field regime limiting the aspect ratio increase.

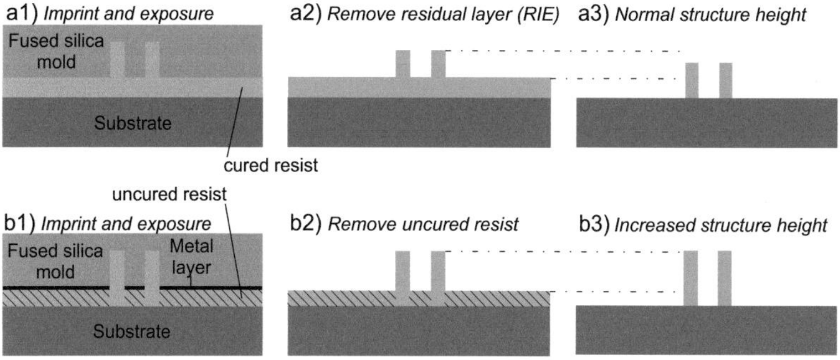

Figure 3.6: *Increased aspect ratio due to hybrid nanoimprint molds.*

3.3.4 Residual Polymer Layer

The thin polymer layer between mold protrusions and substrate is the residual layer (cf. Fig. 3.3). It is one of the most important optimization parameters for NIL and differentiates NIL from other molding and embossing techniques used as surface topography technique so far. When the imprint mold is pressed into the imprint material, the imprint material under the mold protrusions is squeezed and moves towards mold cavities. The mold cavities are filled first. Once all cavities are filled, the imprint material has to travel large distances to the mold periphery. The movement of the mold is slowed down tremendously due to the reduced effective pressure on the imprint material [24, 84]. By filling the cavities and squeezing excessive material to the mold periphery, the volume of the imprint material between mold and substrate is reduced and the mold approaches the substrate. For useful imprints, this residual layer must be minimized. Due to the preservation of the resist volume during imprinting, the residual layer can be optimized with the initial layer thickness according to the mold layout. However, is cannot be avoided completely in standard imprint processes and without special measures.

The polymer film thickness h_f under a mold protrusion depends on the applied pressure p, the distance between two mold cavities s, the initial resist thickness h_0 and the time, the resist is allowed to flow t_f as long as the cavities surrounding the feature are not completely filled[102]:

$$h_f = \sqrt{\frac{\eta_0 \cdot s^2 \cdot {h_0}^2}{2p \cdot t_f \cdot {h_0}^2 + \eta_0 \cdot s^2}} \tag{3.3}$$

The residual polymer layer has two distinct influences on the optical behavior of waveguides. For neighboring waveguides, the residual layer leads to optical crosstalk [18]. Furthermore, the effective waveguide height is increased and a rib waveguide is produced leading to a modified mode profile of the waveguide: the portion of the evanescent field outside the waveguide is reduced [103]. This portion, however, is crucial for sensor approaches probing the waveguide environment. The acceptable residual layer depends on the certain waveguide geometry and the refractive indices.

Dispensing the resist instead of spin-coating allows to adapt the amount of resist to the feature density on the mold. However, bubble trapping can originate from dispensed resist patterns [104].

3.3.5 Monolithic Molds

A NIL-mold is characterized by its overall patterned area. It determines the imprint behavior. On so-called positive molds less than 50 % of the mold area has protrusions (= a not-recessed or convex regions). A negative mold (cf. Fig. 1.2) has more than 50 % protrusions and, thus, less then 50 % cavities (= concave regions). Direct polymer patterning, e.g., of waveguides requires in principle negative molds. With the help of dummy-fill structures, a positive mold can be achieved. The most important difference in imprint behavior is, that negative molds operate in a cavity-filling mode making the replication more complex [24] than positive-molds operating in a protrusion-embossing mode.

The patterning of NIL molds comprise mostly reactive ion etching (RIE) to realize the required height contrasts. Silicon molds have several advantages over fused silica regarding the RIE process. The silica RIE-process and etch profile control is generally considered to be more challenging compared to silicon RIE. Silica etching is less prevalent in industry and research than silicon etching. The tool base is mainly focused on silicon processing. Thus, a broader experience is present and high resolution etching processes with a high structure fidelity for silicon are easier accessible. Silica etching requires metal masks to provide a sufficient etch selectivity and thus an additional process step to pattern the metal masks. Furthermore, it was found that silicon RIE enables thinner EBL resist layers, improved etch profiles and improved compatibility with chemical amplified resists and thus shorter EBL writing times compared to silica RIE [105]. A further advantage of silicon molds is the ability to selectively remove ASL-residues with an oxide etch. This is not possible for silica molds.

Attenuation losses due to rough surfaces were identified to be the primary reason for optical losses in polymer waveguides [106]. Besides other losses, surface roughness losses are directly related to the fabrication process. The attenuation loss α due to sidewall roughness is directly related to the core refractive index n_{wg} and the cladding refractive index n_{cl} [16]:

$$\alpha \propto (n_{wg}^2 - n_{cl}^2) \tag{3.4}$$

Thus, a waveguide surrounded by water n_{cl}=1.33 and made of n_{wg}=1.6 (polycarbonate) has about twice as much losses than a waveguide made of n_{wg}=1.46 (polyacrylate). The absolute loss itself depends on the waveguide geometry and the refractive index of core and cladding. For single-mode channel waveguides a loss of about 0.03 dB/cm for a roughness of 40 nm was given as a rough orientation [16].

The bottom of RIE patterned structures is considered to be smooth compared to the sidewalls. The main contributions to roughness losses originate from the sidewall roughness. It is known that NIL replicates even the tiniest structures like the mold-sidewall roughness profile "one-to-one" into the imprint material. To avoid additional post-imprint treatments, the sidewall roughness of the mold should be as low as possible. Typical sidewall effects in RIE are scallops due to multi-step RIE processes [107] or vertical, chamfer-like structures (see Fig. 3.7). The sidewall roughness can be improved by adequate RIE-process parameters like the kinetic energy of ions. However, optimization of the sidewall roughness has complex interdependencies with other RIE parameters and etch results [107].

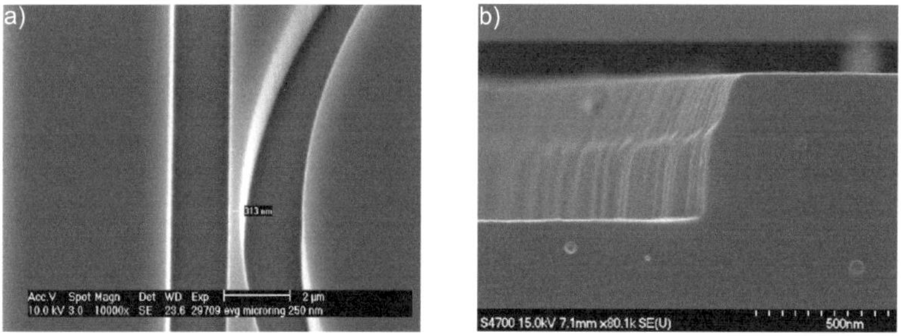

Figure 3.7: *Vertical chamfer structures in silicon-RIE for a) a microring resonator and b) close-up for another trench sidewall (step in the sidewall profile was due to a discontinuity in in the RIE process).*

A sidewall roughness of approximately 10 nm can be achieved for silicon RIE with optimized processes [108, 109]. Post-etch smoothening of silicon sidewalls by wet chemicals enable even lower RMS of about 6 nm [110]. Another group used thermal oxidation and a subsequent oxide back-etch to reduce the RMS roughness from 30 nm to 6 nm after oxide removal [18]. However, in this case the dimensional change was about 300 nm (per edge and on the bottom). For RIE of silica, optimized roughnesses of about 20 nm were reported [111]. The roughness for low-loss silicon waveguides have to be in the order below 10 nm [112] being close to the process limits. Due to the lower refractive index, the RMS values for polymer waveguides are somewhat relaxed and should be in the order below 30-40 nm [16, 106, 110]. For microring resonators, due to changed mode profiles in bended structures even lower sidewall roughnesses smaller than 10 nm are required for very high quality resonators [106].

This demonstrates that an additional pretreatment of either the etched NIL-mold or the imprinted structure is required to reduce the sidewall roughness further below 10 nm RMS, which is even more difficult for silica than for silicon molds.

3.3.6 Bonded Molds

Automatic nanoimprinting, mold-cleaning and tooling require a certain, standardized mold size and geometry. Especially stamp-and-repeat imprinting often requires small active imprint areas. Manual imprinting uses mainly customized systems, however, even therefore imprinting, cleaning and complete handling of small structures and molds without pattern destruction is required. The handling of a mold with a large support is much simpler than for example that of a single, structured element with dimensions of, e.g., $10 \times 10\,\mathrm{mm}^2$ and a thickness of less than 1 mm. An alternative to the single "small-chip"-approach are bonded nanoimprint molds (BIMs).

In contrast to monolithic molds which are made from one material without joints (see Fig. 3.8a), BIMs are assembled by mounting a patterned chip, bearing the imprint structure, to a base support. In this way, also different materials can be joined and non-monolithic molds can be achieved. Based on the discussed intramolecular forces and adhesion models (cf. Secs.2.1, 2.4), different methods can be used to join bonding partners. BIMs can be assembled via direct bonding methods (see Fig. 3.8b) or via mediated bonding methods (see Fig. 3.8b). In principle, direct bonding methods as anodic bonding or high temperature, plasma assisted and wet chemical assisted bonding are known from wafer bonding.

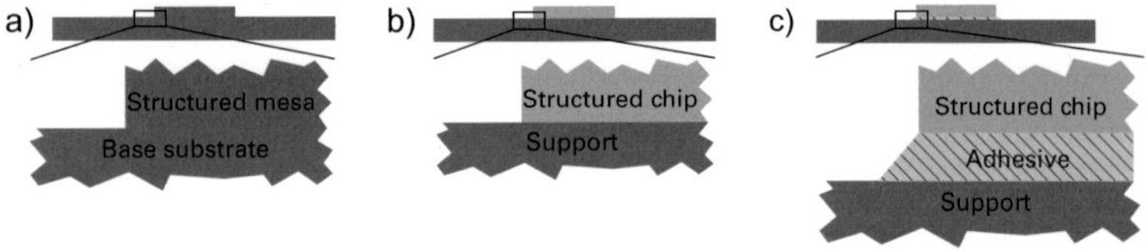

Figure 3.8: *Comparison of a) a monolithic imprint mold with a patterned mesa and bonded imprint molds realized either by b) direct bonding methods or c) adhesive bonding.*

Anodic Bonding This is a common technique in wafer bonding based on an induced ion drift and resulting electrostatic forces. Anodic bonding can be applied to silicon and any kind of alkali-oxide rich glass. The bonding partners are placed on each other and pressed together by two electrodes. A high voltage is applied between both electrodes, resulting in a net movement of the positive alkali cations in the glass towards the negative electrode and away from the bonding interphase. The cations leave oxygen anions behind. The silicon is positively charged relative to the glass due to the electrodes. The resulting net electrostatic force across the interface pulls both surfaces further together and results in a drift of oxygen into the now narrow silicon-glass interphase producing permanent silicon-oxide bonds on the interface between glass and silicon. The movement of the ions can be supported by additional heating. [113]

Anodic bonding joins materials of different coefficients of thermal expansion resulting in interface tensions upon temperature variations. The presence of alkali ions is a limitation when anodic bonding of BIM should be used in the front-end of electronics manufacturing as would be the case for NIL. Compared to direct bonding, anodic bonding can be achieved with surfaces up to 1 µm surface roughness [114].

Direct Bonding Direct bonding of silicon or silica surfaces depend on hydrophilic (hydroxyl terminated) or hydrophilic (hydrogen terminated) surfaces. A surface hydroxylation can be achieved by by wet chemicals like acidic piranha, diluted sulphuric acid and nitric acid or by oxygen plasma processes. Hydrophilic, oxidic, particle and contamination free surfaces with low thickness variations, a high flatness and smooth surfaces bond spontaneously when brought in intimate contact due to hydrogen bonds between surface hydroxyls and superficial water of both bonding partners. Typical wafer flatness variations in the order of 1-3 µm are acceptable. The typical roughness for wafer bonding should be in the order of 0.5 nm RMS. The initial bonding strength in terms of the interface energy after contacting at room temperature is in the order of 100-150 mJ/m^2. Storage and additional annealing results in increased interfacial energies of more than 2 J/m^2. During annealing the water is removed from the interface and covalent siloxane (Si-O-Si) bonds are formed. Several low temperature approaches exist to minimize the thermal impact on the substrates. [115]

Hydrophobic surfaces are prepared by removing oxide layers from silicon surfaces by etching in diluted and buffered hydrogen fluoride. Upon intimate contacting at room temperature, spontaneous bonding occurs due to van-der-Waals interaction of the hydrogen terminated surfaces. Initial interfacial energies are in the order of 20-30 mJ/m^2. A thermal treatment leads to a hydrogen desorption for temperatures larger than 400 °C and the formation of covalent silicon (Si-Si) bonds. [115]

A wet chemical- and plasma-assisted direct bonding was recently applied for a bonded UV-NIL mold [116]. Direct bonding works without an additional material layer and enables thus a high planarity between chip and support. But direct bonding requires additional equipment, is more complex and difficult to integrate into automatic handling systems than mediated-bonding methods.

Adhesive Bonding Whilst direct bonding of silicon or fused silica chips to a base support requires well polished bonding partners, mediated bonding can be achieved for rougher surfaces. A fast and flexible, mediated-bonding method is adhesive bonding (cf. Fig. 3.8b). It can be done with standard equipment like flip-chip-bonders, stamp-and-repeat imprint systems or even simple laboratory parallel-plate pressing setups. Further, a direct bonded joint will fail for high temperature applications due different coefficients of thermal expansion and stress between the chip and the support. The stress results in delamination of less stable bound chip-support systems. Adhesive bonding is thus highly suited for T-NIL, where the adhesive joint can partly compensate different coefficients of thermal expansion of the chip (e.g. silicon) and the support (e.g. fused silica or silicon carbide). A hotmelt was used to bond fused silica chips in a reusable principle to a glass support [71, 117].

Besides organic adhesives also inorganic adhesives like glass-frits can be used. Mediator free techniques directly deploy the adhesion of surfaces to each other.

The chip can be structured with standard semiconductor wafer-equipment. This enables a cost-efficient batch-processing of many chips on a single wafer. For the final BIM the single chips are diced from the wafer. The support of a BIM should be thick enough to provide mechanical support and prevent distortion under high imprint forces. Chip and support are normally transparent for UV-NIL. For T-NIL, opaque materials like silicon carbide and silicon can be used for example as support and as chips, respectively. For NT-UV-NIL only a transparent support or a transparent layer on the substrate is needed. The chips can be of any other, also opaque material (cf. Sec. 4.4). Bonded nanoimprinting molds are an alternative to expensive monolithic molds, but can be used with the same imprint equipment as monolithic molds.

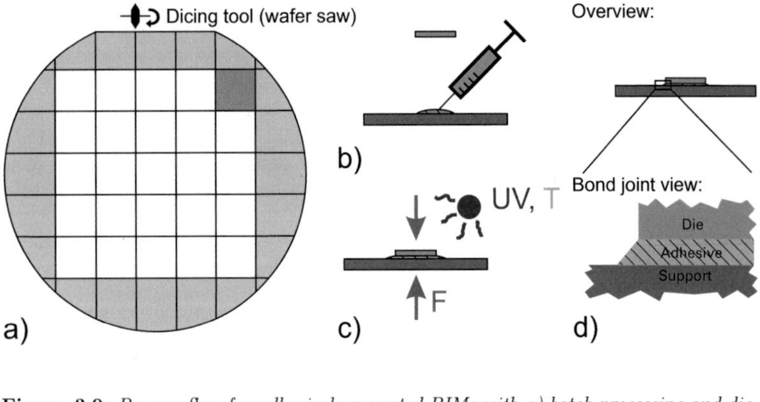

Figure 3.9: *Process flow for adhesively mounted BIMs with a) batch processing and dicing of structured chips from a wafer, b) dispensing of the adhesive, c) curing of the adhesive and d) the final BIM material stack.*

3.4 Functional Surfaces and NIL

Surface functionality of a surface arise from both topography and chemistry. As already mentioned, NIL is a dedicated polymer patterning technique and enables the detailed replication of arbitrary topographical features down to sub-50 nm. Chemical functionalities can be added to such already topographical functionalized surfaces either by

- post-imprint treatment or
- incorporation of functional additives into imprint materials.

3.4.1 Functional Groups for Bioconjugation

Certain chemical functionalities are required to attach biomolecules to a surface. The MRR-structures realized in this work are intended for the specific detection of target molecules or analytes. Therefore, biomolecules with specific binding or affinity sites have to be immobilized on the MRR polymer surface. Several concepts exist to attach biomolecules to such a polymer surface, which is also referred to as bioconjugation. A good overview on this topic is given by *Goddard* [118]. The subsequent section is mainly based on this overview.

The most important strategies for bioconjugation are

- electrostatic adsorption,
- ligand - receptor pairing and
- covalent attachment.

Electrostatic adsorption uses the forces of opposite charged molecules and surfaces to attach a particular molecule to the surface (see Fig. 3.10a). Ligand and receptor pairing like the biotin-avidin linkage uses a unique recognition and binding of the biotinylated sites to avidin or streptavidin (see Fig. 3.10b). The biotin-avidin linkage is one of the strongest non-convalent bonds for bioconjugation. Avidin was reported to bind unspecifically to hydrophobic surfaces and to provide binding sites for biotinylated molecules [17]. The most favorable attachment is realized by covalent bonding (see Fig. 3.10c and d). It offers the most stable bioconjugation bond (cf. also 2.1).

It is possible to directly bind the target biomolecule to the polymer surface or to use spacers or so-called tether molecules. This allows to tune the distance between the surface and the biomolecule, the modification due to the used functional groups (universal group on the substrate and different tether functionalities) as well as the density of the available functional groups (multi-functional tethers). Typical groups for bioconjugation are

- thiols (-SH),
- aldehydes (-CHO),
- carboxylic acids (-COOH),

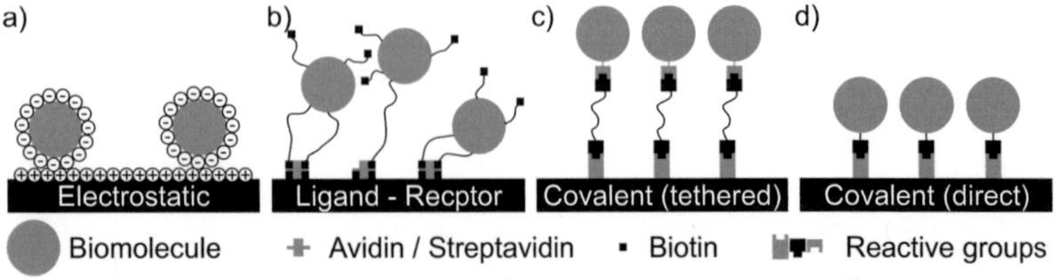

Figure 3.10: *Immobilization techniques for bioconjugation using a) electrostatic forces, b) the affinity of certain ligands and receptors, and c)-d) the covalent binding of reactive groups. Sometimes c) tether molecules are used to optimize the immobilization (modified from [118]).*

- hydroxyls (-OH) or
- primary amines ($-NH_2$).

A uniform distribution of defined and specific groups is required for an efficient bioconjugation. Techniques which result in a random distribution of diverse groups are in this context less useful.

3.4.2 Post-imprint Treatment

An advantage of post-treatment is that the imprint polymer complexity could be kept low because no functional additives have to incorporated into the imprint polymer. Typical techniques for post-imprint polymer modification are

- wet chemical treatment,
- silane coupling agents,
- UV-irradiation,
- ionized gas treatment.

Wet chemical modifications can be easily done in simple laboratory setups and enable an efficient modification of 3d and porous materials. Unfortunately, these treatments are not very specific and results in different functional groups. Wet chemical modifications results depend also on the treated polymer itself. Silanes are mostly used for inorganic substrates (cf. Sec. 2.8), but can also be used to modify hydroxylated polymer surfaces. Therefore, an additional hydroxylation step of the polymer is required if other groups than hydroxyls are present after the functionalization step. Laser functionalization under certain gas environment is also applied [119]. It is known that UV-exposure can yield reactive sites, which can be changed to functional groups after exposure to a particular gas atmosphere. Ionized gas treatment is mainly done by plasma processes.

Other methods like corona discharge or flame treatment can be also applied but mainly yield non-specific functional groups. The desired functionalities can be influenced by the plasma environment and the process conditions. The spectrum of available functionalities in semiconductor compatible processes is mostly limited to oxygen, nitrogen or fluorine containing functionalities due to the typically available semiconductor process gases. Oxygen plasma treatment can be used for oxygen containing groups. Carbon dioxide is useful for carboxyl groups. Amine groups can be realized with plasma processes using ammonia or nitrogen as process gas. Also inert gases like argon can be used to generate radicals on the polymer surface, which can be used for further modification. Thiols are more difficult to obtain. Wet or plasma treatments can be also used to activate the surface for subsequent treatments like silanization.

Wet treatment can selectively etch polymer components and roughen the surface. This is especially a problem for polymer waveguides, as it results in higher propagation losses. Wet chemicals might further results in structure delamination if the process conditions are too harsh. A plasma modification can be done for many different polymers. The post-treatment of polymer surfaces might change the surface topography. For example, selective etching of different polymer constituents can induce surface roughness. A topographical change might also alter the desired device functionality. A broad spectrum of functional groups might result from plasma treatment. It is possible to change functional groups from one type to another and to homogenize functional groups with particular reagents. This implies additional material treatments with unwanted side-effects.

One challenge for polymer functionalization by the named techniques is that only the superficial material layer is modified. Due to the tendency of the polymer to reduce the interfacial energy to the environment and to increase the bulk polymer entropy, functional groups can migrate into the bulk. These groups are lost for further use. Crosslinking helps to minimize this migration. Thus, polymer synthesis using a high amount of crosslinkers are quite suitable for functional polymers. Furthermore, if the functionalities are homogeneously distributed in the bulk material, the migration will be further minimized.

3.4.3 Incorporation of Functional Additives

For direct imprinting of microring resonators, UV-NIL and low-viscous imprint materials were preferred in this work. The following sections are thus focused on incorporating functional additives into UV-NIL imprint materials. UV-nanoimprinting of functional materials was recently reviewed by *Ofir*. This section introduces the basic principles. Additional literature and further details can be found in his work [120].

Functional imprint materials can be used in a wide variety of applications. Direct patterned functional materials can be used, e.g., in optical or electrical devices like waveguides or organic electronics. Topography and chemical functionalities could also influence the behavior of biological cells. Certain superficial chemical groups can be exploited to start particular reactions from the imprinted surface enabling for example the grafting of other functionalities to the surface.

Functional materials can be grouped into i) intrinsically functional materials and ii) non-functional matrices with incorporated fillers or functionalities. Intrinsically functional or active materials are for example semiconductive polymers [121], the more biological relevant polyethylene glycol [122] and chitosan [123]. Examples for filler materials are Ag-nanoparticles [124] or carbon nanotubes [125], which can enhance the electrical conductivity of a non-conductive polymer matrix. As an advantage of the latter one, an already optimized and well understood imprint material (see Section 3.4.4) can be used as matrix material. This reduces the effort of optimizing imprint process parameters. This is a very frequently seen approach for chemically functional imprint materials. It was shown that the desired functionalities withstand the imprint process and are available after imprinting without degradation [120, 123].

For example, the non-functional pre-polymer formulation consists of free radical curing monomers and a photoinitiator. A bulk monomer is used, which makes up the greatest number of the repeating units in the final polymer network [126]. This could be a mono-functional monomer having a single reactive or polymerizing group. To enhance the mechanical stability, often a crosslinker connects different growing chains of the bulk monomer [126]. Crosslinker have more than one polymerizable group. The photoinitiator generates reactive radicals upon UV-exposure and starts the polymerization. Dedicated fillers can be dispersed in this non-functional material to raise certain functionalities. By adding radically polymerizable functional additives the desired funtionality is bound to the polymer matrix and "locked in place". After releasing the imprint mold, the functionalities are part of the bulk polymer properties and superficial groups are available for further surface reactions. [120]

If a modification of the surface chemistry could be excluded, even a plasma etching of the material without loosing the functionality can be done as proposed by [120] because "fresh" functional groups are exposed after removing some material from the surface.

In summary, the incorporation of additives increases the imprint polymer complexity but enables a much broader spectrum of functionalities ranging from simple chemical functionalities and functionalities for bioconjugation, to electrical or optical active components. Imprinting functional materials reduces the required process steps as an additional post-treatment is omitted. NIL enables both the topographical and the functional engineering in the same process step. Functional imprint materials can be thermoplastic materials for T-NIL or UV-curing materials for UV-NIL. Hybrid approaches like thermal curing of low-viscous pre-polymer solutions [127] or initial UV-curing followed by a subsequent final thermal curing [126] are possible.

3.4.4 UV-NIL Material Requirements

As discussed in Section 3.3, the imprint polymer viscosity is the most eminent difference between T-NIL and UV-NIL. The primary requirements for a UV-NIL imprint material are

- a low viscosity,
- a high photo-sensitivity,
- a good adhesion to the substrate and a minimal adhesion to the mold.

For binary patterning, the viscosity should be below 5 mPa·s. This makes the UV-NIL process robust against pattern density variations and minimizes the mold damage due to a lubricated mold-substrate interface. It also allows an accurate in-situ mold alignment to the substrate. Low viscous materials can be applied with manual and automated dispense systems. [92]

Secondary material requirements as

- a low vapor pressure,
- good wetting of mold and substrate,
- no phase-segregation,
- a high mechanical strength and high energy dissipation upon demolding as well as
- a low shrinkage.

improve the imprinting results and simplify the process control. Non volatile substances are required to minimize the change in composition during imprinting [92]. A good wetting of the imprint molds is important to fill high aspect ratio structures and small features as well as to minimize air entrapment. Simulations revealed, that the contact angle with both the substrate and the mold has to be optimized to minimize air inclusions [128]. Very important for functional imprint materials is the homogeneous blending of functional additives in a basic formulation [92]. A phase-separation would lead to inhomogeneous polymer properties. Mechanical properties like Young's modulus, elongation at break and break stress are important for demolding and subsequent use of the imprinted features. The cohesion of the material and a sufficient dissipation of the demolding force is especially important for high aspect ratio features and a optimal feature integrity [92, 126]. A good thermal stability is required for metal evaporation and lift-off. The shrinkage of UV-NIL materials impacts the pattern quality. On one hand, shrunken features enable a better demolding due to reduced friction at the sidewalls. Especially for high aspect ratios as for microring resonators, this could improve the imprint result. On the other hand it leads to a dimensional off-set and to structure deformation.

Typical resist formulations for binary UV-NIL are based on acrylates. An alternative material class are vinyl ethers. Acrylates are the most widely used material class. Acrylates offer a low viscosity and a quick radical curing upon UV exposure. Very important for functional material development is the wide variety of commercially available acrylate monomers. Vinyl ethers have even a lower viscosity than acrylates. However, they have higher vapor pressures than the comparable acrylates, result in a higher release force and their commercial availability is limited. [126]

A large drawback of acrylates is there sensitivity to oxygen [129]. Oxygen inhibits the polymer chain growth and slows down or even completely prevents the resist curing. Oxygen inhibition can be partly overcome by the exposure dose and the initiator concentration. When the initiator concentration is reduced below a critical limit, the degree of curing will be also reduced because the termination reactions during polymer chain-growth and cross-linking will dominate the process. The oxygen is depleted during the resist curing. If the oxygen diffusion into the resist is slower than the oxygen consumption, the resist is fully cured. During imprint, the imprinted resist is covered by the imprint mold. Oxygen can only diffuse into the resist from the mold periphery. Thus, the available oxygen is limited and the imprinted feature is fully cured. However, the excess material, which is squeezed out from under the mold forms a so called edge bead at the mold periphery. This wall is in direct contact with atmospheric oxygen and needs higher curing doses to be fully cured. This increases the processing time or could lead to a structure contamination with uncured resist (see Fig. 5.20). This effect can be minimized by flushing the imprint area with inert gases like nitrogen, argon or helium to minimize the oxygen inhibition effect. Vinyl ethers not oxygen sensitive [126, 129].

One problem of inorganic functional additives is the template cleaning. Almost all organic residues can be removed from the mold by simple solvent rinsing or more advanced by acidic piranha or oxygen reactive ion etching. Inorganic compounds are difficult to remove in this way. For example silicon containing residues are difficult to be selectively removed from a fused silica mold due the conversion of silicon to silica during an oxidizing cleaning procedures [126]. There are other concepts like the use of reversible cross-linkers to remove inorganic resist residues [126].

Experimental Results

4 Technology Components for UV-NIL

"Anybody who has been seriously engaged in scientific work of any kind realizes that over the entrance to the gates of the temple of science are written the words: 'Ye must have faith.'"

(Max Planck)

This chapter describes the basic components required for the manufacturing of chemically functional microring resonators.

First, the developed chemical vapor coating system for ASLs is discussed. For these ALSs, Angle resolved XPS measurements revealed a thickness of about 0.8-1.2 nm and indicated dense, amorphous silane layers. The chemical stability of silanes against a combined attack with acids and oxidation agents was not completely clear from the literature and detailed investigations were made for acidic piranha. The achieved chemical stability against piranha was in the same range as the stability during UV-NIL. The UV-NIL stability in terms of SFE increase was comparable high to other published data. Al_2O_3 was introduced as alternative antisticking layer.

Secondly, the synthesis of a hydroxyl-functional imprint material is demonstrated. The chemical accessibility and tuning of the functional density was proofed with a specific fluorine tag and unpatterned surfaces.

Thirdly, adhesively bonded imprint molds were established as alternative for expensive molds during process development, imprint material evaluation and less residual layer critical techniques like lift-off patterning.

Finally, a technique based on bonded imprint molds with opaque chips was developed for direct polymer patterning and lift-off on opaque substrates with transparent top layers in the micron range. The technique is based on the indirect resist exposure from diffuse light.

4.1 Antisticking Layers

Due to its superior coating characteristics (cf. Sec. 2.8), chemical vapor phase deposition was used for all NIL-molds within this work.

4.1.1 FDTS-CVD Setup

A system for CVD coating of silanes was developed together with *N. Tham* [130]. The system consisted mainly of a borosilicate glass cylinder with 85 mm outer diameter and 3.5 mm wall thickness. Variable cylinder lengths can be used for different amounts of samples and different sample sizes. The system was optimized for $65 \times 65\,mm^2$ UV-NIL standard molds with a thickness of 6.35 mm. A single-chip coating was also possible. The cylinder was sealed with two highly heat resistant silicone sealing caps (see Fig. 4.1a). A temperature sensor and vacuum port were inserted through one cap. The complete setup was self-sealing during evacuation and a laboratory vacuum pump stabilizes the absolute pressure at about 30 mbar.

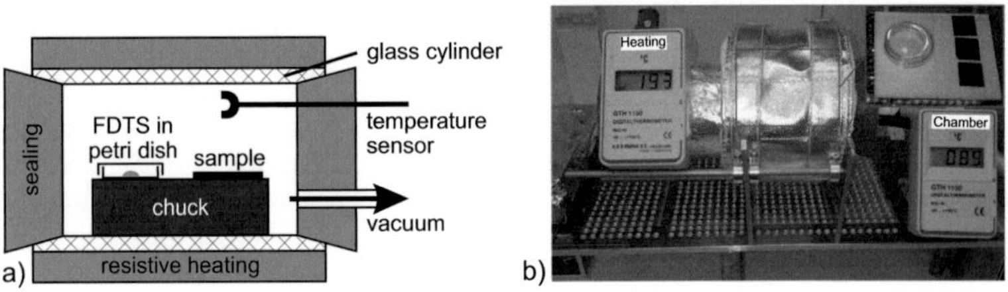

Figure 4.1: *a) Schematic and b) real representation of the developed chemical vapor phase deposition system.*

A typical coating regime started with the temperature setting. The accuracy of the temperature including all errors was about ±2.5 °C. Once the complete chamber reached the coating temperature, the activated specimen was placed in the chamber together with a certain amount of FDTS in a petri dish. After closing, the complete chamber was evacuated and held under constant pressure. Only during loading and during the first 30 s of the coating regime, the temperature was below the desired temperature due to the loading. The FDTS was manually pipetted in a covered petri dish to avoid spillings of FDTS on the samples, which was often seen for non-covered dishes due to the rapid evacuation of the chamber. The aim of the chamber was to evaporate FDTS, which was achieved by a low pressure and a sufficient temperature. Comparable high evaporation rates could be also achieved by either purely thermal driven or pressure driven evaporation. This would either require a temperature of about 220 °C or an ambient pressure of about 0.01 mbar[1]. The goal was to reduce the coating temperature by using lower pressures. On one hand, this reduced the temperature impact on the sample. For example

[1]Estimated with Clausius-Clayperon relation from Tab. 2.5.

hydroxyl groups coalesce at higher temperatures and might be then not available for an FDTS reaction with the surface. In normal atmosphere, temperatures above 200 °C are required therefore. At reduced pressures, this temperature might be also reduced. On the other hand, room temperature evaporation for trichlorosilanes is not suitable due to the lower volatility compared to alkoxy-silanes citeJanssen2006 but were preferred in this work due to their higher reactivity. The reduced pressure also minimized the amount of water being present in the chamber atmosphere. Furthermore, the probability for silanes to react with each other in the chamber atmosphere was also reduced. Both together were assumed to increase the antisticking layer quality compared to normal pressure evaporation. The amount of used FDTS differed for individual samples but was always in the excessive range of 70-140 µl FDTS leaving unconsumed FDTS in the petri dish at the end of the procedure. Before a new glass chamber was used for the first time, the interior was pre-conditioned by a FDTS coating without samples to ensure a minimal FDTS-consumption and reproducible results during sample coating. Even if the FDTS-precursor was stored for a long time and first particulates were visible, these particulates were never vaporized and ensured very reproducible coating results.

Already the first coating results demonstrated the much higher quality of a vapor phase process compared to compared to liquid phase processes (see Fig. 4.2). After only one minute of liquid coating, the surface was covered with large particulates (see Fig. 4.2a). After 30 min large crystalline, multilayer OTS-islands grew on the surface (see Fig. 4.2b). In contrast to this, even after prolonged vapor phase coating only a few particulates were visible. No difference in the roughness was observed between moderate and long vapor coating regimes up to 30 min.

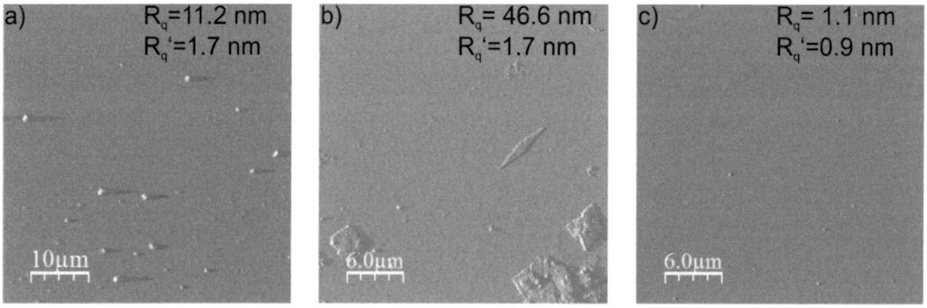

Figure 4.2: *Atomic force microscopies of a) liquid phase coated OTS (10 mM in toluol, 1 min dip), b) liquid phase coated OTS (10 mM in toluol, 30 min dip) and c) vapor phase coated FDTS (≈20 µl, 120 °C, 30 min) [130]. Inserted data indicate the RMS-roughness of the complete sampling area (R_q) and the particulate free area (R_q').*

To ensure a homogeneous FDTS-layer, a stable binding and high amount of cross-links (cf. Sec 2.9.3) to the substrate, the coated surfaces were tempered for five minutes at 150 °C and subsequently for one minute at 250 °C. The temperature was ramped-up in less than ten minutes to the according temperatures and cooled down slowly at the end to avoid abrupt temperature steps.

4.1.2 Properties of FDTS-CVD Layers

Initial Surface Free Energy Different typical NIL-mold materials were coated with the CVD-procedure describe below. Native oxide, thermally grown oxide and fused silica surfaces were quite comparable after and only partly before FDTS-CVD-coating. The SFE-spread over all experiments was larger before than after coating. The reason was most probably differences in the pre-treatment of the substrates before coating. After coating, a very low SFE with a central value around 15 mJ/m^2 was obtained (see Fig. 4.3). The relative narrow spread indicated a reproducible SFE even for different coating temperatures. In contrast, the chromium surface CrA (chromium surface type A) seemed to be more sensitive to variations in the coating parameters, especially regarding the temperature. The smaller SFE of uncoated CrA-surfaces originated from the fact that it was a non-SiO_2 surface and had different properties compared to the silica surfaces. Compared to liquid coating, the vapor process yielded a similar low SFE and thus a dense and high quality FDTS layer.

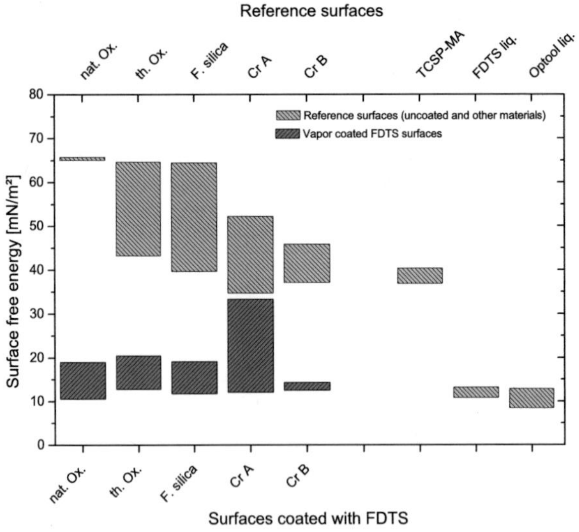

Figure 4.3: *Overview on min-max-spread of all SFE-measurements made during the course of this work for different substrates and coatings.*

To find the optimal coating process, the coating temperature was varied. The SFE after coating was used as quality criterion. The coating pressure and time was fixed. It was proofed that the coating process will depend on the first few minutes [63]. A prolonged coating did not influence the final SFE significantly. Thus, a fixed time of 15 min was chosen.

Even though the absolute SFE values were different (see Fig. 4.4[2]), the optimum coating temperature for all materials was about 75-100 °C. This was close to the FDTS boiling point at 30 mbar (cf. Tab. 2.5). The coating at 74, 100 and 145 °C was done for all materials and each temperature in the same process-run. A clear increase of the SFE with the temperature was present. Similar results were obtained by other groups [131]. Additional experiments with other temperatures confirmed that finding. Possible reasons for higher SFE at larger temperatures are a reduced amount of reactive hydroxyl sites or surface water. The same effect was observed by another group and increased desorption of silanes at higher temperatures was mentioned as possible reason therefor [56]. Only for CrA very high temperatures were tested. The fact that the SFE decreased again for very high temperatures might be attributed to an increased evaporation rate of FDTS. In any case, the FDTS-kinetics seemed to change with increasing temperature. The addition of water to the coating environment (volume ratio water to FDTS was 1:2) had no significant influence on the SFE-temperature relation for higher temperatures. For 80 °C the SFE was increased. Results from other experimenters [63, 130, 132] were used to complement the obtained results. CrA, an antireflective chromium-oxynitride, had in all cases a higher SFE than the other oxidic surfaces. The most reproducible results were obtained for the standard UV-NIL material fused silica. Due to different coating parameters the absolute SFE value differed between different experimenters for the other materials. As a consequence, the optimal coating regime was chosen to be 15 min at 75-100 °C without additional water for all imprint molds.

Thickness and Coverage The measurement and modeling was done by *A. Kleiner* [63]. It is important to understand, that thickness and coverage are interdependent: If one of both is constrained during the simulation, the other will be adjusted by the optimization algorithm to get a low RMS value. For example, if a 50 % coverage is preset, a respective layer thickness d will be found. If a 100 % coverage is preset, the thickness will be reduced to $0.5 \cdot d$ to achieve a good fit. The theoretical thickness of an FDTS-layer is 1.4-1.6 nm [55, 133, 134]. However, lower experimental values were frequently reported due to partially formed monolayers, tilted molecules [133] and an amorphous FDTS-layer [55].

Results: Uncoated fused silica surfaces had a 0.2-0.4 nm thin contamination layer with a coverage of 50-100 %. This can be, e.g., organic adsorbants or functional groups like hydroxyls. For modeling of the different layers (cf. Sec. 2.9.2), firstly the theoretical

[2] Differences in SFE values compared to earlier publications arise from the attempt to use a unique set of test liquid properties throughout all contact angle measurements made during the course of this work.

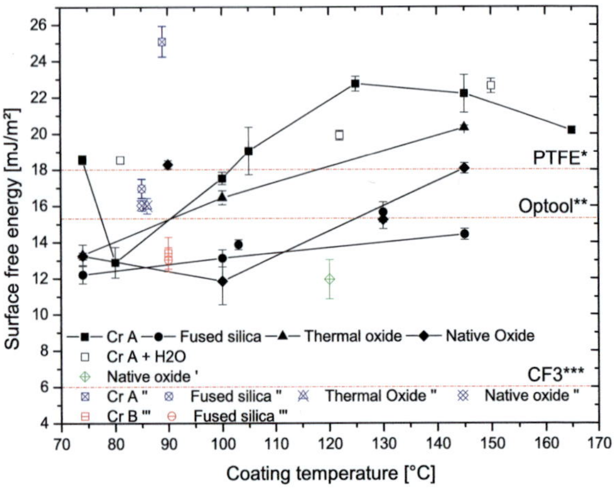

Figure 4.4: *Comparison of the SFE for different coating temperatures and different substrate materials including data from '[130], "[132] and "'[63]. (*,*** [33], ** [117]) Note that the results are specific for the used setup.*

stoichiometric composition of the FDTS, the substrate and a $C_6H_6O_2$-contamination was used. Additionally, the solver was allowed to modify these compositions. This reflected the possibility that the contamination on an FDTS coated sample can differ from that on fused silica. Additionally, a non-ideal, modified FTDS and substrate composition can be considered by this approach. In the following models, the resulting spread for the obtained values is given as measure for the variation of the results.

Three different basic models for an FDTS-layer and a possible contamination layer on fused silica were compared (see Tab. 4.1). First, a partially covering FDTS-layer on a fully covering contamination layer was assumed (model 1). In this case the FDTS coverage F_{FDTS} was found to be almost complete. The thickness of the FDTS-layer was calculated to be 0.8 nm. The thickness of the contamination was comparable to the uncoated substrate (see Tab. 4.1, model 1). Secondly, again an incomplete FDTS-coverage was assumed, but with the contamination layer being localized in the non-FDTS covered regions (model 2). These boundary conditions resulted in a partial FDTS coverage. The contamination was quite thick because it was modeled to be completely in the non-FDTS-covered region and not under or on the FDTS (see Tab. 4.1, model 2). Finally, a completely covering FDTS layer was assumed with a contamination partially covering the FDTS (model 3). The FDTS thickness agreed with the values determined so far, but the contamination was comparable thick due to the partial covering (see Tab. 4.1, model 3).

Table 4.1: *Basic AR-XPS models and results for a vapor coated FDTS-layer and a contamination layer on fused silica.*

Model	Modelled layer stack	F_{FDTS} [%]	d_{FDTS} [nm]	F_c [%]	d_c [nm]
1 RMS: 0.6-1.6	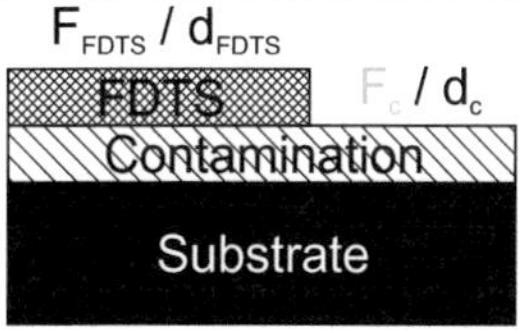	99-100	0.8	100	0.2-0.3
2 RMS: 0.6-1.5	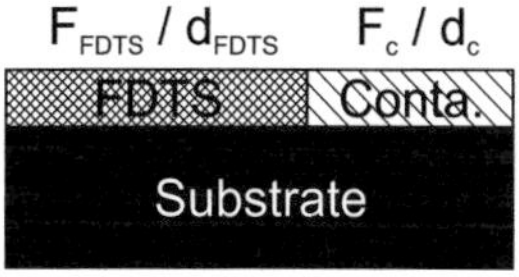	89-90	0.9-1.0	10-11	2.0-2.6
3 RMS: 0.7-1.6	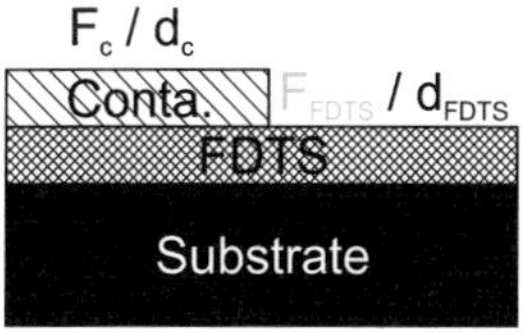	100	0.9	8-12	1.6-5.6

Further models were used to analyze the structure of the FDTS-layer. First, regions with different FDTS-thicknesses were assumed. In further attempts, the FDTS-layer was separated in two layers: a tail-group layer (only CF_3-groups) and a head-group and spacer layer (remainder FDTS parts). The average FDTS-thickness of these models was larger (1.0-1.2 nm) than the before obtained values, but no further data on the FDTS-structure was obtained.

Interpretation: In model 1, the contamination described mostly the interface between the FDTS and the fused silica substrate due to its low thickness. Models 1 to 3 were equal in their RMS-value. However, model 1 and 3 seemed to be more realistic due to the thick contamination in model 2. In model 2, all existing groups which were not included in the FDTS or substrate model were included in the contamination and forced to a limited region. For example, hydroxyl groups may be present at the FDTS and the substrate and also within the FDTS-layer (cf. Fig. 2.8). As a conclusion, the realized FDTS layers had a high coverage of above 90 % and a thickness of about 0.8-1.2 nm. No conclusion on the presence of islands could be drawn from the 90 % coverage. It rather means that 10 % of the surface cannot be explained with the FDTS and can be homogeneously distributed within the FDTS-layer. The thickness was lower than the theoretical value of FDTS and a large thickness variation was found. Both originated from the modelling and the averaging measurements on amorphous silane layers.

Ellipsometry measurements revealed a thickness of about 0.6 nm for the vapor phase deposited FDTS-layers. The SFE for these layers was in the order of 18 mJ/cm^2. *Jung* found the layer thickness after one cycle of their FOTS vapor phase deposition process to be in the order of 0.6-0.7 nm [56]. After three cycles their layer thickness increased but did not exceed 0.9-1.0 nm. They attributed their increase to water added to the coating atmosphere and thus a denser and thicker FOTS-layer. The FTDS vapor phase process used in this work did not operate in a cycle mode. No water vapor could was applied to the chamber to yield new reactive sites. However, it should be noted, that the influence of water depends strongly on the coating setup and procedure. Thus, the thicknesses found by AR-XPS and ellipsometry agreed well with other published data.

As final conclusion, the obtained results did not rebut the quasi-monolayer concept (cf. Sec. 2.8).

4.1.3 Stability of FDTS-CVD Layers

To evaluate the ASL-stability in cooperation with *B. Lu*, $15 \times 15\,\text{mm}^2$ fused silica chips were diced from a 6 in wafer and $15 \times 15\,\text{mm}^2$ chips were diced from a 5 in photomask blank with an antireflective chromium coating (Toppan Printing) termed CrA in this work. XPS of the antireflective chromium revealed a mixed chromium-oxynitride surface (see Tab. 4.2). The C-signal originated from typical organic surface contaminations. Traces of S and Na as well as Cl probably originated from piranha cleaning left-overs and other contaminations. The ASL-quality was evaluated by determining the SFE

Table 4.2: *Atomic composition of the antireflective chromium-oxynitride surface after multiple piranha cleanings.*

C [at.%]	**N** [at.%]	**O** [at.%]	Na [at.%]	S [at.%]	Cl [at.%]	**Cr** [at.%]
21.1	8.7	50.9	0.4	3.0	0.2	15.6

via contact angle measurements with the test liquids deionized water (DI-water) and diiodomethane (DIM) (see Tab. 4.3). The SFE was calculated from the measured contact angles. FDTS is a perfluorinated silane and a high quality ASL thereof should have a

Table 4.3: *Surface tension and its components of the used test liquids for SFE determination (Dataphysics Conctact Angle System OCA10, Database).*

	Surface tension γ_l [mN/m]	Disperse part γ_l^d [mN/m]	Polar part γ_l^p [mN/m]
DI-water	72.10	19.90	52.20
DIM	50.00	47.40	2.60

SFE of about 15-17 mN/m for fused silica and 18-22 mN/m for the CrA surface for the used setup as discussed above. A higher SFE will indicate a degraded ASL. Oxidic surfaces can be activated by plasma or piranha treatment, which results in an increased SFE due to an increased density of hydroxyl groups. However, such high hydroxyl group densities are not stable and decrease until a stable equilibrium is reached. This can be seen in a SFE recovery over time to a lower and stable value. If such a recovery is observed for an investigated surface, this indicates a clear degradation of the ASL and a surface being at least partially oxidic and free of ASL molecules. In general, the standard deviation of the determined SFEs for all treatments was quite large compared to the SFE average. This deviation resulted from sample preparation, FDTS coating and contact angle measurements as well as inhomogeneous degradation on the sample surfaces. The deviation was a measure for SFE differences both on individual samples and between samples of the same batch. However, conclusions on systematic degradation were mainly drawn from the averaged SFE values.

Stability During Cleaning The ASL-stability was tested against the following cleaning procedures:

- Acetone ultrasonic bath (removal of minor and solvable resist residues)
- Acidic piranha (oxidation of major organic resist remnants)
- Oxygen RIE (oxidation and sputtering of major organic resist remnants and removal of other minor residues)

The acetone cleaning was conducted in a petri dish, which was placed in an ultrasonic bath for 10 min. Acidic piranha was prepared by mixing two volume parts of 96 % sulfuric acid (H_2SO_4, VLSI grade) with one part 31 % hydrogen peroxide (H_2O_2, VLSI grade) in a laboratory glass vessel. The initial temperature, when placing the samples, was 100 °C. The mixture was left to react with the samples and gently stirred to ensure a homogenous mixture. Finally, all samples were rinsed in DI-water and blown dry by N_2. The acetone and piranha cleaning was repeated several times to check for long term effects. The RIE was done with an Oxford Plasmalab 80 PLUS (Oxford Instruments). The process parameters were 60 sccm O_2 gas flow, 75 mTorr chamber pressure and 67 W HF input power resulting in a self bias voltage of about 400 V. The treatment time was 60 s per sample.

Acetone: The acetone cleaning had neither a significant degradation effect on the ASL on the fused silica nor that on the CrA surface (see Fig. 4.5a). These findings matched the observed solvent resistance of trichloro-silanes (cf. Sec. 2.9.3). Acetone is a mild cleaning agent and dissolves or swells polymers. It does not attack covalent bonds and, hence, could be used without problems to remove minor resist remnants from imprint molds.

Piranha: Piranha cleaning led in general to a significant increase in the SFE with an increasing number of cleanings (see Fig. 4.5b). Two particular observations were made. Firstly, the ASL on the CrA was completely removed after only one cleaning. The very high SFE of 65.6 mN/m directly measured after cleaning and the recovery to 50.0 mN/m after five days of storage indicated an oxidic surface without FDTS. Second, the SFE was continuously increasing on fused silica. But up to eight piranha cleanings no clear SFE recovery was visible and, thus, the ASL was not significantly degraded. After 12 cleanings a clear recovery and, hence, degradation was visible. The SFE increased with about 1.2 mN/m and 0.9 mN/m per treatment, measured for a period of 20 cleanings both directly after imprinting and after five days of storage.

The observed stability against piranha matches the literature results on oxidation and acid resistance of triisocyanate-silanes as well as the acid resistance of trichloro-silanes on silicon oxides (cf. Sec. 2.9.3). Acidolytic attack was also discussed in literature (cf. Sec. 2.9.3), which could explain the minor but steady degradation of the ASL on fused silica. As a result thereof, a short cleaning with piranha is possible for fused silica surfaces without an intensive ASL degradation.

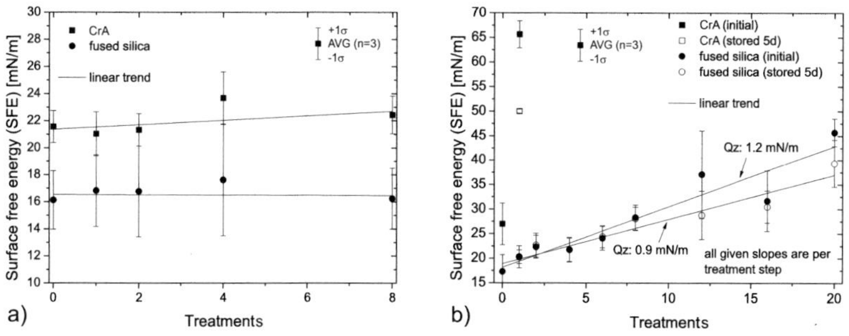

Figure 4.5: *a) Stability of FDTS-ASLs on fused silica and CrA surfaces against multiple acetone treatments (each treatment 10 min) and b) ASL-degradation on fused silica and CrA surfaces due to multiple piranha treatments (each treatment 10 min, "initial" indicates measurements directly after treatment).*

An intensive gas bubble formation was observed when an FDTS coated sample was immersed in fresh piranha. It is known that piranha gradually oxidizes organic materials and produces carbon dioxide thereby. As discussed, the Si-O-Si bond is hydrolyzed under basic conditions (cf. Sec. 2.9.3). Sulfuric acid dehydrates hydrocarbons. Thus, a direct attack of the sulfuric acid on the FDTS alkyl chain seems to be probable. It was assumed that the gas bubbles are carbon dioxide as a byproduct of the ASL oxidation. It was further observed that a short dip into piranha can remove superficial particles from already ASL-coated imprint molds. It was concluded that the produced gas bubbles can trap small particles and physically remove them from the surface. This could help to remove minor particles and contaminations from ASL-coated mold surfaces.

An important aspect was that the ASL on CrA had a quite low SFE before piranha treatment, but was not very chemically stable. This demonstrated that the SFE cannot be used alone to determine the quality of an ASL. At the moment, there is no validated model for the reduced stability of the silane layers on CrA. Proposed explanations are

- a less stable FDTS binding to chromium,
- a less dense or less ordered FDTS layer, or
- a less dense anchoring of the FDTS layer to the surface

compared to silica. The binding stability of silanes to oxides depends on the oxide itself [135]. Thus, the FDTS might have been less stable bound to chromium oxides than to silicon oxides. The SFE after FDTS coating was generally higher on chromium than on silica. The reason therefore can be a less dense or a less ordered FDTS layer. A surface of pure CF_3 groups has a lower SFE than one with mixed CF_2 or CH_2 groups or even an uncoated region (cf. Tab. 2.4, Sec. 2.9.1). Judging from the SFE, the FDTS coated chromium might have exposed more CF_2, CH_2 or uncoated regions than the coated silica. Consequently, potentially piranha sensitive parts of the FDTS-ASL might have been more accessible on chromium than on silica. *Wassermann* observed an inhomogeneous removal of trichloro-silanes and assumed that the degradation starts from defects in the monolayer [67]. Such defects might have been more pronounced in the ASL on chromium than on silica and might gave access to piranha sensitive parts like the alkyl chain of FDTS (see Fig. 2.6). Another possibility was that the ASL was less often anchored to the antireflective chromium surface than to the silica surface due to differences in the hydroxyl group density and morphology. Crosslinking is still under discussion as contribution to monolayer stability (cf. Sec. 2.9.1). As a result of reduced cross-links, the ASL could have been less stable bound to the chromium. A combination of those discussed explanations was assumed and further research is required to evaluate the individual contributions.

Reactive ion etching RIE completely removed the ASL from both the fused silica and the CrA (see Fig. 4.6). Thus, RIE is a suitable method to remove ASL remnants from a mold before recoating a fresh ASL on it. RIE probably destroyed the ASL mechanically due to sputtering and removed organic residues by oxidation.

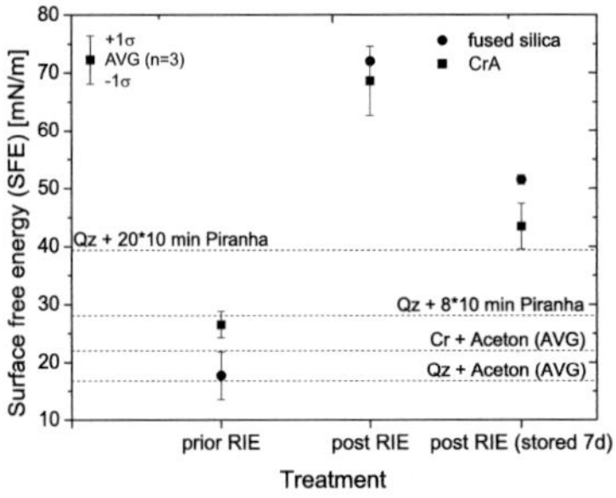

Figure 4.6: *ASL removal from fused silica and CrA surfaces due to a single RIE treatment.*

Stability During UV-NIL The degradation of the samples due to UV-NIL was evaluated using a customized, UV-curing, acrylate based resist system (see Fig. 4.7a and Sec. 4.2). The sample was placed with the ASL facing upwards in a manual laboratory imprint press (see Fig. 4.7b). A roughly 0.5 µl resist droplet was manually dispensed on the sample. A cleaned microscopy glass slide was manually placed on the resist droplet. The complete stack was pressed and UV-exposed for 120 s with about 50 mW/cm^2 (measured at 365 nm). After curing, the stack was unloaded, the samples were manually demolded and the ASL mold surface was inspected for resist residues. Minor residues were removed as long as possible by acetone ultrasonic cleaning. Only persistent residues had to be cleaned with acidic piranha, which had an additional impact on the ASL degradation. The SFE was determined every 10 imprints.

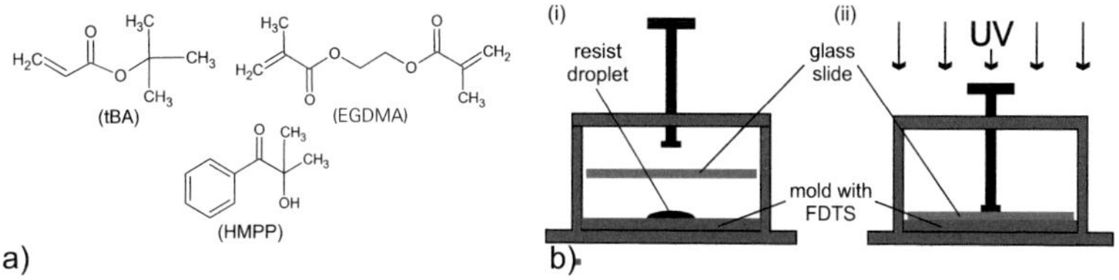

Figure 4.7: *a) Customized resist system for evaluation of the ASL degradation during UV-NIL: (tBA) tert-butyl acrylate as base monomer and reactive solvent: 67.5 wt.%, (EGDMA) ethylene glycol dimethacrylate as cross-linker: 30 wt.%, and (HMPP) 2-hydroxy-2-methyl-propiophenone as tphotoinitiator: 2.5 wt.%. b) Schematic of the manual imprint setup during manual (i) resist application and (ii) UV-NIL imprinting.*

A clear increase of the SFE of 1.2 and 2.2 mN/m per 10 imprints was observed for the fused silica and the CrA surfaces, respectively (see Fig. 4.8). This demonstrated, as already observed during piranha cleaning, that the ASL on CrA was less stable than on fused silica. For fused silica, the observed SFE increase was relatively low and comparable to literature values (see Tab. 4.4).

After 40 imprints with CrA surfaces and 50 imprints with fused silica surfaces, large resist remnants were sticking to the surface and could not be removed by acetone. After removing the resist remnants with acidic piranha, the ASL on the CrA mold was completely degraded (see Fig. 4.8, step for imprint #40), and the SFE on the silica was also increased significantly (see Fig. 4.8, step for imprint #50). The same silica samples were used for further imprints. After ten subsequent imprints, resist residues had to be removed again by piranha. The SFE standard deviation for SFE measurements after the imprints #50, #60, and #70 was larger than that before imprint #50. This indicated also an inhomogeneous ASL surface both on individual samples and between different samples of the same batch.

A steady SFE increase of 2.5 mN/m per ten imprints and one piranha cleaning was observed for silica surfaces after being stored for five days. This was comparable to the sum of the before observed value of 1.2 mN/m per 10 imprints and 0.9 mN/m per single piranha treatment. It evidenced a superposition of both degradation effects. Thus, silica molds imprinted several times cannot be cleaned with acidic piranha without an intensive ASL degradation.

During UV-NIL imprints, first polymer sticking to the mold occurred around SFE values of 22 mN/m. It was, hence, expected that a SFE of about 20-25 mN/m was the maximum SFE for a working ASL with the used setup and materials. However, this is no universally valid value. Sticking to the imprint mold depends on the resist adhesion both to the substrate and to the mold. The higher the SFE of the mold, the larger is the resist

Table 4.4: *Comparison of literature values for ASL degradation per 10 subsequent imprints.*

Slope in [mN/m]	Imprints	Resist	Note	Ref.
1.87	0-250	PAK-01	Optool DSX (liquid phase), UV-NIL	[71]
≈0	0-500	NIF-1-A	Optool DSX (liquid phase), UV-NIL	[71]
1.26	0-189	PAK-01	F13-TCS (liquid phase), UV-NIL	[71]
10.34	1-40	customized, diacrylate	Optool DSX (liquid phase), UV-NIL	[73]
6.02	10-40	customized, diacrylate (+ 3 % F13-TMS)	Optool DSX (liquid phase), UV-NIL	[73]
1.16	20-40	customized, diacrylate (+ 3 % F13-TES)	Optool DSX (liquid phase), UV-NIL	[73]
≈33	0-15	customized, acrylate	Optool DSX (liquid phase), UV-NIL	[136]
0.31	15-135	customized, acrylate	Optool DSX (liquid phase), UV-NIL	[136]
3.39	0-104	customized, acrylate (+ surfactant)	Optool DSX (liquid phase), UV-NIL	[136]
≈26	0-15	customized, acrylate	Optool DSX (liquid phase)	[70]
≈0	0-25	PAK-01	Aquaphobe CF + PFP gas supply	[77]

TMS: trimethoxysilane, TES: triethoxysilane , PFP: pentafluoropropane

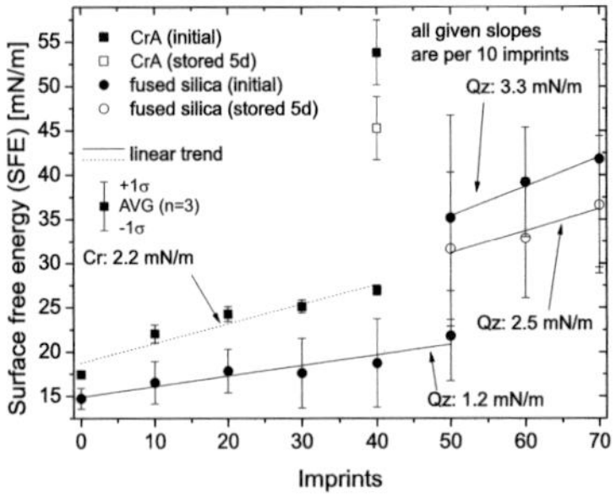

Figure 4.8: *ASL degradation on fused silica and CrA surfaces due to multiple subsequent imprints. Note: The step in the SFE at the imprints #40 for CrA and #50 for silica samples was due to a required piranha mold cleaning.*

adhesion to the mold. A degradation of the ASL increases always the SFE and, thus, the adhesion to the mold. Optimizing the adhesion to the substrate, for example with adhesion promoters, will enable successful imprints with ASLs with even higher SFE. However, the maximum useful adhesion is determined by the cohesion of the resist. *Xu* reported a durability of an FDTS layer on fused silica of about 100 imprints for a similar resist formulation [92], which is quite comparable to the results obtained here without special adhesion promoters.

4.1.4 Aluminum Oxide as Antisticking Layer

Aluminum oxide was tested in cooperation with *L. Teng* [137] as antisticking layer due to its low reactivity and its beneficial surface energy constitution (cf. Secs. 2.4.5 and 2.7.3). The used customized acrylate based imprint resist was comparable to PMMA and a low adhesion to aluminum oxide was expected and experimentally proofed (see Fig. 4.9).

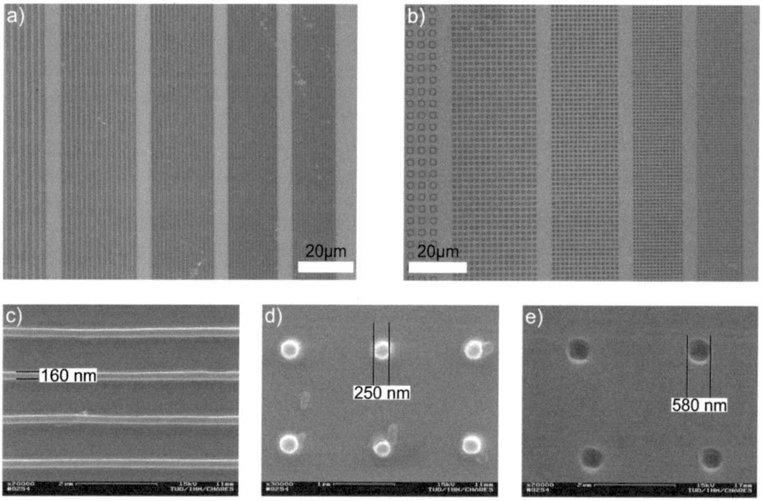

Figure 4.9: *Imprints made with aluminum as ASL in a customized resist formulation on fused silica substrates using TCSP-MA as adhesion promoter: a-b) large imprint fields with sub-500 nm features replicated without adhesion defects and c-e) several small resolved features of these imprints (modified from [137]). Note: Defects in a) originated from mold defects and not adhesion failures.*

A roughly 50 nm Al layer was sputtered on a fused silica mold (PH350, NTT-AT) which was bonded to a glass support (cf. Sec. 4.3). The mold was imprinted in the customized acrylate resist system (cf. Fig. 4.7a) on a fused silica substrate to allow a resist curing by NT-UV-NIL (cf. Sec. 4.4). Later, the sputtered aluminum layer was found to be not completely dense, especially on the feature sidewalls. A partial curing originated from light transmitted through this incomplete aluminum layer. The mold was manually separated from the substrate and no significant defects were found. This principal experiment demonstrated the usability of aluminum as antisticking layer.

An important challenge was the chemical stability of the aluminum ASL. After sputtering, a comparable thick, superficial oxide develops on the aluminum and protects it from further oxidation. However, this oxides was found to be not stable enough to protect the aluminum coating from being attacked during piranha cleaning. The sputtered Al was easily removed in acidic piranha. To check whether aluminum under the superficial Al_2O_3 was oxidized or completely removed together with the superficial Al_2O_3, Al_2O_3-layers of several thickness were deposited by atomic layer deposition on silicon substrates. Subsequently, the samples were cleaned in acidic piranha for 10 min (100 °C initial temperature). The removal of Al_2O_3 was determined by XPS measurements. For Al_2O_3 coated surfaces, no Si signal should be detectable from the substrate because the thinnest Al_2O_3-layer was 10 nm and thus thicker than the maximum XPS sampling depth. This could be proofed for all three layers (see Tab. 4.5). After the piranha treatment, only on the 10 nm sample a significant Si content could be found. However, Al was still detectable to a high extent on the surface. Most of the oxygen could be explained with the Al_2O_3. The remaining part can be almost completely explained by hydrocarbon contamination. Only the 10 nm sample after piranha cleaning differed slightly from that explanation: a small but still notable amount of oxygen cannot be explained in this way and was, hence, assigned to the underlying silicon substrate with an oxidized surface. As a conclusion, Al_2O_3-layer was only partly stable against the piranha cleaning. It had become either thinner than 5 nm which was roughly the maximum XPS-sampling depth or it was still 10 nm thick and had become porous. In any case, parts of the Al_2O_3-layer have been removed. No degradation of thicker layers could be observed. It was assumed that they were also degraded, but the remaining layer was still thicker than the maximum XPS sampling depth.

Table 4.5: *Relative atomic concentration of untreated and acidic piranha treated Al_2O_3 determined by XPS.*

Al_2O_3 Sample	C [at%]	**O [at%]**	F [at%]	**Al [at%]**	**Si [at%]**
10 nm	14.7	57.4	1.3	26.6	0.0
10 nm + piranha*	18.8	51.7	1.3	18.7	9.6
20 nm	16.1	56.1	0.3	27.6	0.0
20 nm + piranha	12.4	56.7	1.8	29.1	0.0
30 nm	13.5	57.3	1.2	28.0	0.0
30 nm + piranha	14.4	57.1	0.0	28.5	0.0

* measured and averaged at two independent sites

The oxide layer on top of the sputtered aluminum layer was in the order of only 5-10 nm. This was not sufficient for a stable ASL-layer and highly limited the wet cleanability of aluminum ASLs. Thick Al_2O_3 applied for example by atomic layer deposition might be useful as ASL being moderately stable against piranha. As long as the mold is sufficiently cleaned by self cleaning effects during imprinting, aluminum with a thin oxide and Al_2O_3 can be used as ASL. However, both layers will induce dimensional structure modifications. Al_2O_3 has the further advantage to be transparent and does not rely on NT-UV-NIL.

4.2 Hydroxyl-functional Imprint Polymer

The scope was to synthesize a functional imprint material which can selectively bind molecules to the imprinted polymer surface by covalent bonding. As mentioned, hydroxyl groups can be used for bioconjugation (cf. Sec. 3.4.1). This required an imprint material which provided after polymerization accessible hydroxyl groups on its surface. As a vision, such a material can be used to bind a specific antibody to the polymer surface. The specific antibody will recognize for example a virus, which in turn will be bound to the antibody and, thus, to the polymer surface. If the polymer pattern is a microring resonator, a specific virus would be attached to it (see Fig. 4.10) and enable the detection of the virus by an MRR detuning. In this way, a selective biosensor could be achieved.

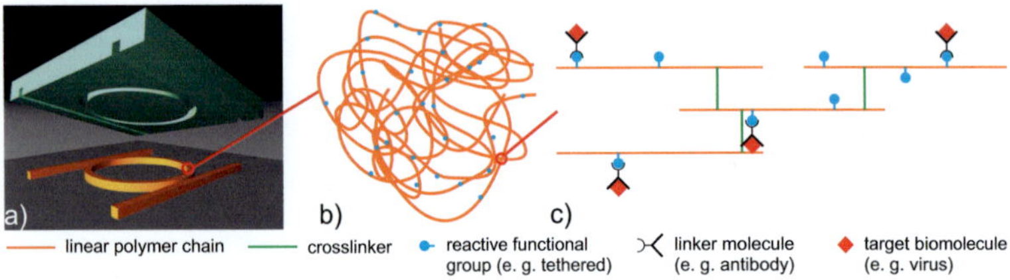

Figure 4.10: *Concept of UV-imprinting of a chemically functionalized microring resonator with a) microring resonator device, b) macroscopic polymer chains and c) details of the polymer chains with functional additives being integrated into the polymer matrix and enabling a bioconjugation with antibodies.*

To directly pattern microring resonators, a negative mold is required (cf. Fig. 4.10a). As discussed before (cf. Sec. 3.3), UV-NIL and low-viscous imprint materials will be best suited therefore. Additionally, UV-curing materials enable hybrid techniques and residual polymer layers free imprints (cf. Sec. 3.3.3). No residual layer influences the chemical functionality of the polymer. Important for a sensor application are further inert polymers with a high conversion rate and a minimal amount of volatile compounds influencing the sensor performance.

4.2.1 Imprint Material Formulation

Due to their low viscosity, the broad experience with them and their commercial availability, acrylates were chosen as material class for the functional imprint material. Parts of an already optimized acrylate imprint formulation [126] was used. Instead of achieving a high dry-etch stability, the purpose was to add the required chemical functionalities for bioconjugation. The silicon containing monomer in the *Long*-formulation was exchanged against the hydroxyl functional monomer 4-hydroxy butyl acrylate (HBA). Thus, the new formulation had no inorganic additives and the established oxidizing cleaning procedures with O_2-RIE and piranha (cf. Sec. 4.1.3) could be applied. The original *Long*-formulation had a homogeneous distribution of the introduced silicon [92] and the same was expected for the added HBA.

All monomers were used as received (Sigma-Aldrich Chemie GmbH) without further modifications. A basic, low molecular weight, UV-curing monomer formulation was obtained with the known *tert*-butyl acrylate (tBA) (98 %) as base monomer and reactive diluent, ethylene glycol dimethacrylate (EGDMA) (tech. grade 90 %) as cross-linker for high mechanical stability and 2-hydroxy-2-methyl-propiophenone (HMPP) (97 %) as photoinitiator (see Fig. 4.11). HMPP, which is also commercially available as Darocur 1173, was chosen because it is liquid and has a high miscibility in organic solvents and acrylates. This enabled an easy incorporation in the monomer formulation. HMPP is further well suited for clear coatings and minimal yellowing upon prolonged UV-exposure. Comparable high concentrations of about 2-4 wt% are recommended for 5-20 µm thick layers. Thicker layers require a lower concentration [138]. The described system underwent a radical co-polymerisation upon UV-irradiation in the 365 nm range. This basic formulation was modified by substituting the tBA with different concentrations of HBA (tech. grade 90 %) ranging from 10 % to 75 % (see Tab. 4.6). The viscosity of the monomer system was about 1 mPa·s.

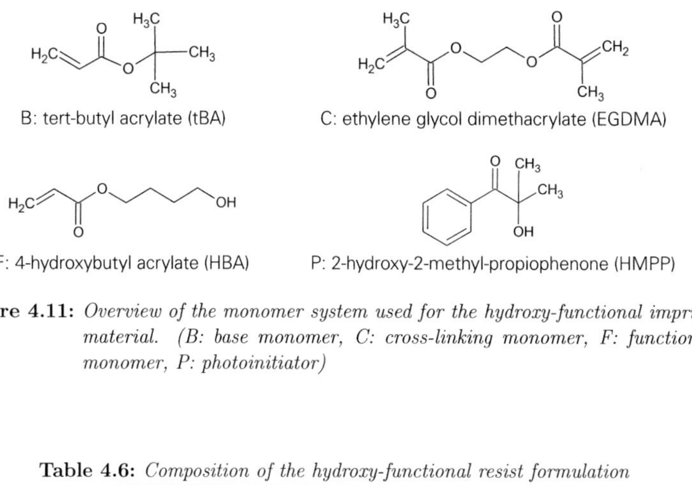

Figure 4.11: *Overview of the monomer system used for the hydroxy-functional imprint material. (B: base monomer, C: cross-linking monomer, F: functional monomer, P: photoinitiator)*

Table 4.6: *Composition of the hydroxy-functional resist formulation*

tBA [wt%]	67.5 - X	HMPP [wt%]	2.5
EGDMA [wt%]	30.0	HBA [wt%]	X=f·67.5
		(f=0.1, 0.25, 0.5, 0.75)	

4.2.2 Specimen Preparation for Characterization

To evaluate the presence and availability of the hydroxyl-groups after curing, specimen were prepared in a three-step process (see Fig. 4.12). First, $10 \times 10\,mm^2$ Si-substrates with 140 nm thermal SiO_2 were cleaned with acetone and isopropanol.

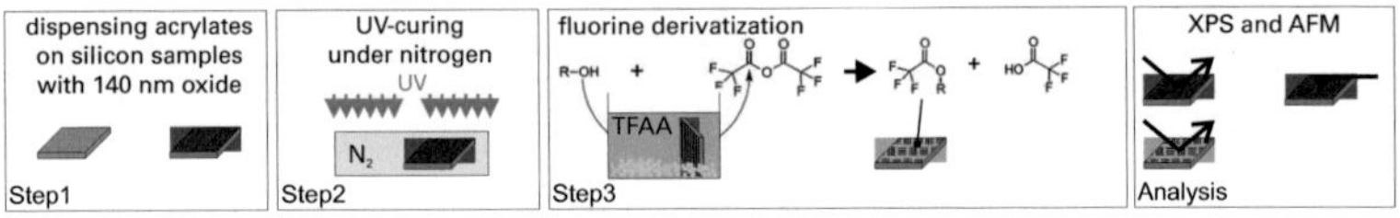

Figure 4.12: *Overview on the specimen preparation and characterization steps for functional polymer surfaces.*

Because all analytical methods required flat and unstructured samples, the monomer mixture was only dispensed on the cleaned substrates and not imprinted. Due to the very low viscosity of the monomer formulations, a large amount of the mixture was dispensed giving a sufficient film thickness to prevent the dewetting of the substrates. The film thickness was about 1-10 µm. Second, all specimen were put into a vacuum and inert gas oven (SRO 701, ATV Technologie GmbH). The oven was purged with a constant nitrogen flow. The resulting chamber pressure was about 200 mbar overpressure. The nitrogen prevented an oxygen inhibition of the radical curing monomers. Curing was induced by a iron-doped mercury arc lamp and a flood exposure. The samples were exposed for about 10 min through a glass window. The intensity on the sample surface was about $25\,mW/cm^2$ (at 365 nm). Due to the large thermal capacity of the oven, the substrate temperature did not exceed 30 °C. Third, after UV-curing, samples were immediately put into the XPS-chamber to prevent environmentally initiated surface reactions. The specimen for chemical derivatized XPS (CD-XPS) (cf. Sec. 4.2.3) were wet-chemically derivatized with fluorine. The derivatization was achieved by placing the samples vertically for 30 min on glass beads in a trifluoroacetic anhydride (TFAA) solution (derivat. grade, $\geq$99 %,Sigma-Aldrich Chemie GmbH) (see Fig. 4.12:Step3). TFAA binds specifically to OH-groups and is a common XPS tag [139]. A delamination of the polymer-films from the SiO_2 during TFAA-tagging was observed. To ensure a sufficient adhesion during tagging, the reactive silane adhesion promoter 3-(trichlorosilyl)propyl methacrylate (TCSPA-MA) (Sigma-Aldrich Chemie GmbH, CAS 7351-61-3) was dip-coated for 20 min in a 1 mM solution in isooctane after a O_2-RIE surface activation. After TCSP-MA coating the samples were cured at 150 and 250 °C. Only with the adhesion promoter, a TFAA-tagging without polymer-film delamination was possible. After derivatization, the samples were also immediately put into the XPS-chamber. Non-derivatized specimens were used for atomic force microscopy (AFM) analysis.

4.2.3 Characterization of Functional Polymer Surfaces

Presence of OH-groups XPS was applied to determine whether the hydroxyl groups introduced by synthesis were present on the surface. The sampling depth of XPS was about 5-10 nm. Thus, only the superficial region of the specimen was probed. The C1s-peak was analysed for contributions by functional oxygen groups using their respective binding energies (see Tab. 4.7).

Table 4.7: *Bindings and binding energies used to determine the presence of superficial OH-groups*

Binding	Binding energy [eV]	Characteristic for ...	Alternative interpretation
C−C	285.0	polymer backbone	-
C−O	286.7	OH-groups	polymer backbone
O=C−O	288.7	polymer backbone	-

The C−O binding signal was used as a measure for the introduced hydroxyl groups. The polymer backbone contributed to a large extend to the amount of C−C bindings and also partly to the amount O=C−O bindings. The C1s XPS-spectra revealed the presence of C−O bindings (see Fig. 4.13) for all monomer formulations. From those spectra, no clear relation between the amount of C−O bindings and the HBA could be seen.

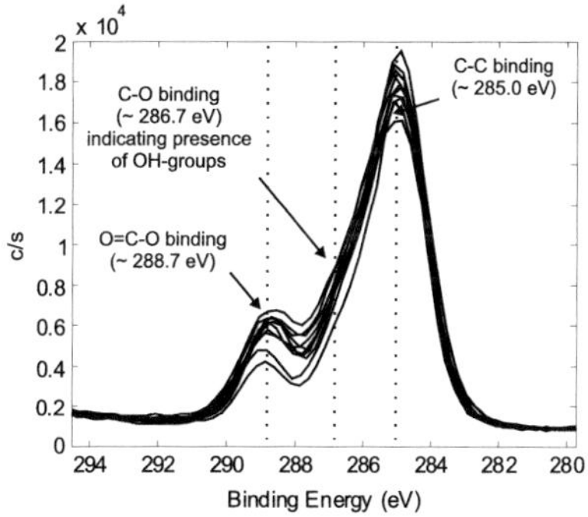

Figure 4.13: *Combined XPS C1s single spectra of surfaces of OH-functionalized polymers and the non-functional formulation indicating the presence of C−O bindings close to or on the polymer surface.*

Unfortunately, the co-polymer backbone contributed also by its molecular structure to the C−O binding signal. Thus, the C−O binding signal was not a specific measure for superficial OH-groups. It was expected, that the C−O contribution from the HBA increases whilst the C−O backbone contribution remains constant. Thus, the overall amount of C−O bindings must increase. To evaluate this theory and to differentiate between OH-groups intentionally introduced into the monomer solution and the backbone C−O bindings, the theoretically expected concentration ratios for C−O, O=C−O and C−C bindings were calculated and compared to measured concentration ratios. The theoretical ratios r_t were calculated from the molecular structures (see Fig. 4.11) and the individual monomer concentration.

$$\begin{aligned} r_{t,1} &= r_{t,CO/CC} \\ &= \frac{n_{C-O}}{n_{C-C}} = \frac{\sum n_{C-O,i} \cdot \chi_{C-O,i}}{\sum n_{C-C,i} \cdot \chi_{C-C,i}} \end{aligned} \tag{4.1}$$

$$\begin{aligned} r_{t,2} &= r_{t,OCO/CC} \\ &= \frac{n_{O=C-O}}{n_{C-C}} = \frac{\sum n_{O=C-O,i} \cdot \chi_{O=C-O,i}}{\sum n_{C-C,i} \cdot \chi_{C-C,i}} \end{aligned} \tag{4.2}$$

$$\begin{aligned} r_{t,3} &= r_{t,OCO/CO} \\ &= \frac{n_{O=C-O}}{n_{C-O}} = \frac{\sum n_{O=C-O,i} \cdot \chi_{O=C-O,i}}{\sum n_{C-O,i} \cdot \chi_{C-O,i}} \end{aligned} \tag{4.3}$$

with $i = (B, C, P, F)$

B, C, P and F represent the base monomer, the cross-linker monomer, the photoinitiator and the functional monomer, respectively. n was the amount of the particular bindings within the four components B, C, P and F (see Tab. 4.8) after polymerization. A complete conversion of C=C bonds to C−C was assumed. Thus, each carbon double bond was counted as one additional carbon single bond. Carbon double bonds within the aromatic ring of the photoinitiator were not separated from carbon single bonds. Deviations from this ideal assumption originate from incomplete conversion and chain termination reactions resulting not in carbon single bonds. χ was the molar fraction of the respective component in the monomer mixture obtained from Tab. 4.6. For example for the 10 % HBA substitution one obtained for χ: 68.96 mol% tBA, 22.04 mol% EGDMA, 2.21 mol% HMPP and 6.81 mol% HBA. This gave with the respective amounts of n (cf. Tab. 4.8) the theoretical value $r_{t,1} \approx 0.190$.

The respective ratios $r_{m,1}$, $r_{m,2}$ and $r_{m,3}$ were calculated from the C1s peak analysis.

$$r_{m,1} = r_{m,CO/CC} = \frac{g_{C-O}}{g_{C-C}} \tag{4.4}$$

$$r_{m,2} = r_{m,OCO/CC} = \frac{g_{O=C-O}}{g_{C-C}} \tag{4.5}$$

$$r_{m,3} = r_{m,OCO/CO} = \frac{g_{O=C-O}}{g_{C-O}} \tag{4.6}$$

g was the relative amount of the respective binding within the C1s peak. For example for the 10 % HBA 16.05 % of the C1s peak was attributed to C−O bindings (g_{C-O}) and

66.01 % to C−C bindings (g_{C-C}). The remainder was attributed to O=C−O bindings ($g_{O=C-O}$). This gave $r_{m,1}$=0.243 which was higher than the above obtained $r_{t,1}$.

The ratios $r_{m,1}$ and $r_{m,3}$ were used as indirect measures for the amount of superficial OH-groups. The value of $r_{m,2}$ was interpreted as a measure for O=C−O rich organic surface contaminations, which are normal on all samples processed under ambient atmosphere. For example, if the $r_{m,2}$ value would have been larger than expected but the $r_{m,1}$ would have been as expected, this was rather attributed to a surface contamination than to the polymer composition because only the $r_{m,2}$ value was influenced.

Table 4.8: *Amount n of particular bindings per monomer component after polymerization. (Note: Cleaving one carbon double bond during polymerization creates one new carbon single bond.*

Monomer component	C−C	C−O	O=C−O
B: tBA	5+1	1	1
C: EGDMA	7+2	2	2
P: HMPP	10+2	1	0
F: HBA	5+1	2	1

If a homogeneous distribution of the OH-groups exists in the polymer bulk as well as at the polymer surface and complete monomer conversion is reached, $r_{t,1}$ and $r_{m,1}$ must be well comparable. Furthermore, both ratios must increase with increasing amounts of HBA because more C−O bindings are incorporated by a HBA concentration increase than are removed by a tBA concentration decrease. Deviations from this ideal assumptions originate from an inhomogeneous material composition, an incomplete conversion and an evaporation of components. The tBA had the lowest vapor pressure of all monomers. It cannot be excluded that tBA evaporated quite fast from the monomer mixture. Measurements revealed an evaporation rate of about 0.67 mg/min under ambient conditions. To account for the influence of the evaporation, r_t ratios were additionally calculated for the extreme situation of a complete tBA evaporation. Because the mentioned influences were difficult to quantify, the trend of the theoretical r_t and experimental r_m ratios was more meaningful than the absolute value.

A continuous increase of the $r_{m,1}$ value up to 50 % tBA substitution was calculated from XPS-measurements for samples without using an adhesion promoter (see Fig. 4.14a). This increase was proportional to $r_{t,1}$. The absolute value was higher than expected. The $r_{m,1}$ increase was attributed to an increased amount of HBA. Due to the limited sampling depth of XPS, an increased amount of OH-groups in the upper 5-10 nm region of the sample surfaces was concluded. The $r_{m,1}$ dropped surprisingly for the 75 % tBA substitution. This evidenced an altered polymer composition for large HBA concentrations. The excess of HBA monomers yielded a surface being depleted of OH-groups. Up to the 50 % tBA substitution, a comparably homogenous co-polymerization could be thus expected.

The $r_{m,2}$ value was much larger than expected (see Fig. 4.14b). Such a large deviation was not observed as strongly for $r_{m,1}$ and for $r_{m,3}$ ratios and was attributed to O=C−O rich surface contaminations rather than the polymer composition.

The trend of $r_{m,3}$ follows the expected decrease with increasing tBA-substitution (see Fig. 4.14c) due to an increased amount of OH-groups close to the surface.

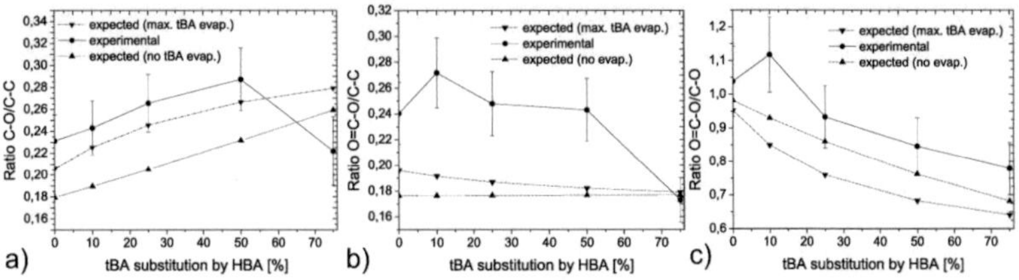

Figure 4.14: *Measured and calculated ratio of a) C−O to C−C, b) O=C−O to C−C, c) O=C−O to C−O bindings*

Accessibility of OH-groups As mentioned, r_m gave an evidence for existing OH-groups on the surface or close to the surface but could not be taken as an absolute measure for the functional group density because the C−O binding was not unique and with the XPS a sum signal of the upper 5-10 nm of the polymer was measured. Thus, also buried OH-groups were measured, which will be not accessible for binding events on the surface. To better quantify the amount of truly accessible OH-groups, TFAA was used as a specific and highly sensitive XPS tag. Each bound TFAA molecule added three fluorine molecules to one OH-group. Therewith, the concentration FD of OH-groups per carbon atom was calculated from the atomic concentrations of carbon (c_C) and fluorine (c_F).

$$FD = \frac{1/3 \cdot c_F}{c_C - 2/3 \cdot c_F} \tag{4.7}$$

Fluorine has a very large photoelectron cross-section. Thus XPS is very sensitive to fluorine and small amounts of fluorine derivatized hydroxyl groups could be measured. The derivatization efficiency of TFAA was unknown but was assumed to be constant. The determined fluorine concentration c_F and the calculated functional densities FD represented a relative measure for the relation between the HBA content and the amount of accessible OH-groups. For an exact calculation of the absolute functional density, more details on the TFAA derivatization reaction would be needed. However, this was beyond the scope of this work.

With increasing tBA substitution, more fluorine was present on the sample surfaces (see Fig. 4.15) confirming an increased amount of TFAA surface reactions. This was attributed to the increased amount of HBA. However, the increase of FD by a factor of

2.2 and 1.4 was not proportional to the respective increase from 10 % to 25 % tBA substitution and from 25 % to 50 % tBA substitution. The reason could have been that the possible tBA substitution, resulting in a homogeneous polymer, approached a physical limit. Or the TFAA-binding was not proportional to the OH-surface density. The calculated functional density *FD* reached a maximum of 8.1 % for the 50 % tBA substitution. For the 75 % substitution no significant amount of fluorine was bound to the surface. This confirmed the already observed drop in the amount of superficial OH-groups for more than 50 % tBA substitution, which seemed to be the physical maximum for the used system. The low fluorine content for 0 % tBA substitution represents an offset to the determined functional density due to naturally occuring OH-groups, other hydroxyls being intrinsically present in the polymer matrix, unspecific adsorption of TFAA or reaction byproducts.

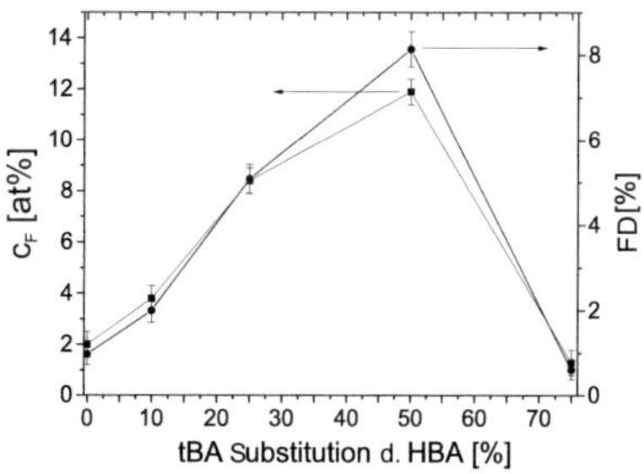

Figure 4.15: *Measured fluorine concentration c_F and calculated functional density FD.*

Topography of Functional Polymer Films Sample surfaces were analyzed via atomic force microscopy (AFM; DI3500, Veeco Instruments Inc.) to determine whether the HBA incorporation influences the surface roughness. No influence on the surface roughness was observed for increasing amounts of HBA. The surface RMS roughness of the samples was between 1.2 nm and 2.2 nm. The differences between different HBA concentrations (see Fig. 4.16) were attributed to a statistical variation of the RMS roughness across the surface. The RMS roughness of an uncoated thermal SiO_2 surface used as reference was about 1.5 nm with the same AFM setup. Thus, no significant roughness differences were measurable for the polymer surfaces compared to this reference.

Compared to typical sidewall roughnesses of nanoimprint molds in the 30 nm (peak-to-peak) range (see Sec. 3.3.5), the intrinsic surface roughness of the functional polymers itself had no influence on the final waveguide characteristics. From that point of view, the functional polymer was well suited for optical waveguides. The final roughness of the directly imprinted waveguides will be mainly determined by the imprint mold rather than the functional polymer.

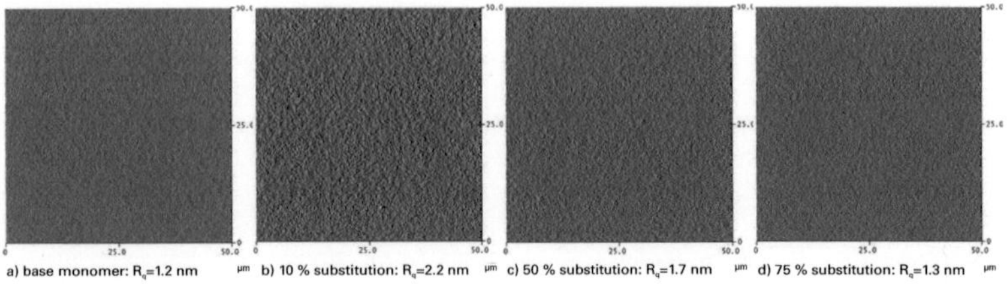

Figure 4.16: *Topography and R_q (RMS) roughness of exemplary samples for different amounts of tBA substitution.*

4.2.4 Outlook for Imprinted Structures

With ellipsometry on µm-thick and non-functional polymer-film a refractive index of about 1.48 was measured at 590 nm. From this measurements, a refractive index of 1.45 to 1.46 at 1550 nm was extrapolated. Further details of the impact of the refractive index are discussed in Sec. 5.1.

The characterization of imprinted functional structures was not performed up to now. The XPS-spot was difficult to position exactly on an imprinted feature. Furthermore, it was too large to measure directly on µm-sized, imprinted structures. It is difficult to predict the behavior of the functional moieties during imprint. However, *Ofir* reported that firstly functional and polymerizable additives, as used in this work, are anchored to the polymer matrix and that secondly enough moieties on the surface exist after imprinting to favor subsequent surface reactions [120]. It was expected, that sufficient hydroxyl groups will be available on the surface after imprint. There amount as well as the correlation with the used tBA concentration have to be verified by experiments. Is is further assumed that the concentration plays a role for the distribution of the moieties in the polymer. First indications for that were seen in the results for high HBA concentrations. It is known that an orientation of liquid molecules on a solid surface can be induced by the solid itself. Especially in adhesion theory, this is thought to play an important role. It is known that such orientation occurs mainly for high energy surfaces [22]. One explanation is the theory that similar force interact with each other (e.g. dispersion forces with dispersion forces). In terms of an FDTS-ASL and an hydroxy functional monomer there is no significant interaction of polar forces possible. The dispersion forces are present in all monomers as well as in the FDTS layer. Thus, there is no effect of phase separation expected as no force-gradient between different monomers exist. The only slightly polar FDTS can exert no force on the monomers. As long as no higher polar forces are present within the polymer or at the substrate, a homogeneous hydroxyl group distribution can be assumed.

4.3 Adhesively Bonded Nanoimprint Molds

Even though polymeric molds posses a great potential for NIL and some companies recently promoted techniques allowing for a conformal imprint on wafer-scale, polymeric imprint structures were not used as systematically as rigid imprint structures for BIMs in this work. The scope was on rigid UV-NIL molds. However, adhesive bonding can be applied universally for all imprint techniques and most of the results are also relevant for NT-UV-NIL and T-NIL. A typical bonded NT-UV-NIL mold as routinely used with the NPS300 imprint system throughout this work is shown in Fig. 4.17.

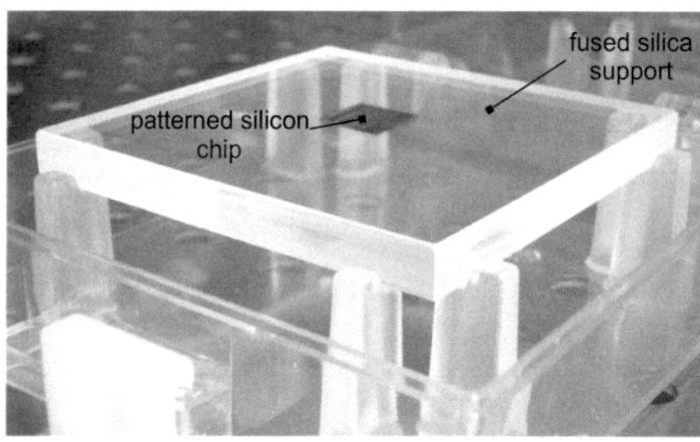

Figure 4.17: *Typical bonded mold composed of an opaque 6.35 mm thick and 65×65 mm^2 wide fused silica support (cf. Sec. 4.4) and a patterned silicon chip diced from a 4 in wafer.*

4.3.1 BIM Requirements and Adhesives

Adhesive BIM can be designed to be demountable [71] or, as discussed in the following, durable for multiple imprints and cleanings. A great practical advantage of durable BIMs was that particles and contaminations from the bonding process can be removed by standard cleaning procedures even with harsh chemicals like acidic piranha. The main requirements on such BIMs for being usable in nanoimprinting are:

- a high bond joint reliability,
- a high surface planarity and
- a long term stability of the bond joint.

As a result of the above mentioned criteria, the adhesive for adhesively bonded imprint molds has to meet the following specifications:

- a high adhesion to the chip and the support (mechanical stability),
- a high cohesion of the resist (mechanical stability),

- a low viscosity (thin bond joints),
- a low shrinkage (low stress and deformations),
- a high chemical resistance (process stability),
- no UV or thermally induced degradation (process stability) and
- a high UV-transmission for UV-NIL (usability).

Typical forces required to detach a 1 cm^2 mold from cured UV-NIL resist were in the order of 10-100 N during all UV-NIL imprints made in this work. The cohesion of the adhesive and the adhesion to the support and the structured chip must be sufficient to prevent a delamination of the chip. A low viscosity enables a thin adhesive joint and thus highly parallel bond joint between the structured chip and the support. Thin bond joints will, e.g., not allow large wedge errors. Parallel joints will further result in planar mold surfaces. Low cure-shrinkage prevents mechanical tensions in the adhesive and warping of the structured chip. The cured adhesive must be highly chemical resistant to enable multiple usages and a high throughput. During the development of new imprint materials and techniques, harsh mold cleaning and processing might be needed and the mold should withstand all these processes without the need for being rebonded.

Different adhesives were screened as bonding materials for BIMs (see Tab. 4.9) in co-operation with *L. Teng* [140]. The samples for these experiments consisted of $10 \times 10\,mm^2$ large and 0.675 mm thick fused silica chips and $20 \times 20\,mm^2$ wide and 2.3 mm thick non-alkali borosilicate glass supports. The chips were bonded with the different adhesives to the supports. The material stack chip-adhesive-support was pressed with the before mentioned parallel-plate setup (cf. Fig. 4.7b) for the adhesive investigations. To meet the above-mentioned specifications, the adhesives were tested for the following properties.

- initial bond strength (via shear force measurement),
- bond joint thickness (with cross sections),
- surface planarity (with cross-sections, laser scanning profilometry),
- chemical resistance (against acetone, oxygen plasma, acidic piranha),
- process stability (against UV and thermally induced aging) and
- usability for imprinting (with different imprint tools).

For later application in stamp-and-repeat imprinting the BIMs were routinely assembled with the NPS300 imprint system (cf. Fig. 4.17).

Table 4.9: *Overview on the different materials screened for adhesively bonded imprint molds.*

Name	Description	Source
	UV-curing adhesives	
A+A	67.5 wt% tert-Butyl acrylate 30.0 wt% Ethylene glycol diacrylate 2.5 wt% 2-Hydroxy-2-methylpropiophenone	Customized acrylate
A+M	67.5 wt% tert-Butyl acrylate 30.0 wt% Ethylene glycol dimethacrylate 2.5 wt% 2-Hydroxy-2-methylpropiophenone	Cust. acrylate-co-methacrylate
M+M	67.5 wt% tert-Butyl methacrylate 30.0 wt% Ethylene glycol dimethacrylate 2.5 wt% 2-Hydroxy-2-methylpropiophenone	Customized methacrylate
Vitralit 4731	Structural epoxy (900-1500 mPa·s)	Panacol-Elosol GmbH
Katiobond 4670	Filler epoxy (4800 mPa·s)	Delo GmbH
	Polymer-solvent adhesives	
PMMA 120k	T-NIL resist Poly(methylmethacrylate) 120 kg/mol	Micro Resist Technology GmbH
PMMA 950k	Electron beam lithography resist Poly(methylmethacrylate) 950 kg/mol	MicroChem Corp.
mrI-T85-1.0	Non-polar T-NIL resist (thickness at 3000 rpm 1 µm)	Micro Resist Technology GmbH
mrI-T85-5.0	Non-polar T-NIL resist (thickness at 3000 rpm 5 µm)	Micro Resist Technology GmbH
PS 143 E	6 wt% polystyrene dissolved in toluene (thickness at 3000 rpm 500 nm µm)	BASF AG
	Thermally curing adhesives	
mrI-9030E	T-NIL resist	Micro Resist Technology GmbH
PDMS	Sylgard 184 (10:1 - base:initiator)	Dow Corning Corp.
Araldite 2020	Optical, two-comp. epoxy (150 mPa·s)	Panacol-Elosol GmbH
EpoTek 301-2	Optical, two-comp. epoxy (225-425 mPa·s)	Epoxy Technology Inc.
EpoTek 305	Optical, two-comp. epoxy (150-250 mPa·s)	Epoxy Technology Inc
	Adhesion promoter	
APTES	3-Aminopropyl-triethoxysilane	Aldrich GmbH
TCSP-MA	3-Trichlorosilyl-propyl methacrylate	Aldrich GmbH

4.3.2 Initial Bond Joint Strength

The initial bond strength was a direct measure for the quality and the strength of the adhesive joint. The initial bond strength can be reduced during the mold-lifetime due to mold cleaning, antisticking layer coating and imprinting. A low bond strength will result in failed imprints due to delamination of the structured chips. Initial bond strengths were measured directly after curing of the adhesive with a shearing setup (see Fig. 4.18). The BIM was placed on a vacuum chuck with an additional mechanical holder. The shearing chisel was pressed against the chip at a constant speed. The chisel height was 500 µm above the support surface. The break force at which the chip was sheared from the support was recorded and limited to about 1 kN.

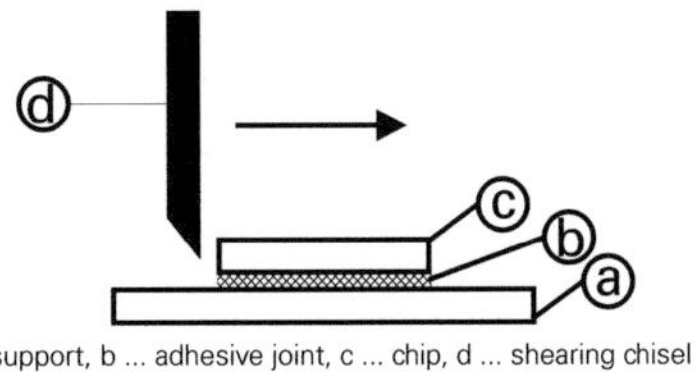

Figure 4.18: *Schematic of the shearing setup*

Particular adhesives revealed shearing tensions of more than 10 MPa (see Fig. 4.19). Except the polymer solvent systems, the other adhesives reached shearing tensions of more than 1 MPa. In comparison, the above mentioned experimental demolding forces of about 10-100 N for an imprint mold with a 1 cm^2 imprint area equals to a pulling tension of less than 1 MPa. Even if the achieved high-strength bonds would degrade to a minor extent, the bond strength would be still more than sufficient to be used in nanoimprinting. The initial bonding strength did not limit the application of BIMs.

Polymer-solvent based systems were baked on a hotplate at temperatures of 60-120 °C for 1 h depending on the polymer. These systems suffered from an ineffective solvent removal from the bond joint and had consequently low bond strengths. Another issue was reproducibility. PDMS for example had an acceptable bond strength but a poor reproducibility (large variance). One reason might have been the large PDMS viscosity resulting in unreproducible bond joints. The best overall results where achieved with UV- and thermal-curing adhesive systems. The curing resulted in higher adhesive forces compared to polymer solvent systems. The latter one depended further more only on physisorption the bonding partners whereas in curing systems reactive species were present which can chemisorb to the respective interfaces. Also a weak boundary layer might have been present in the polymer-solvent system due to remaining solvents (cf. Sec. 2.4). Excellent adhesion was achieved for the customized acrylates A+A, A+M, M+M (cf. Tab. 4.9) by covalently grafting the reactive adhesion promoter 3-(trichlorosilyl)propyl methacrylate (TCSP-MA) (cf. Sec. 2.6) onto both bonding partners. The grafting was

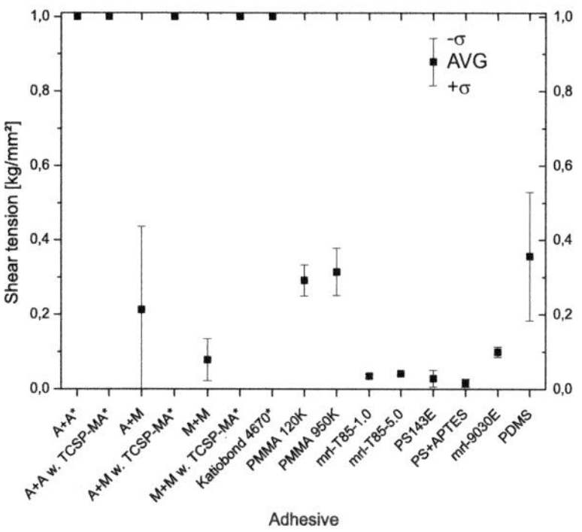

Figure 4.19: *Initial bond strength of all screened adhesive materials directly after bonding expressed as shearing tension. (* measuring limit exceeded)*

done by a 10 min dip-coat in a 1 mM TCSP-MA solution in isooctane and a subsequent bake for 5 min at 160 °C.

The bond strength F_{bond} of adhesively bonded molds arised from two contributions. From an area proportional component F_{area} and from a circumference proportional component F_{circ}. F_{area} depended on the adhesion between chip and support and the contact area between both. If the adhesive joint would be an ideal cubic volume with a lateral extension not larger than the chip, F_{bond} would only depend on F_{area}. In reality, excessive adhesive is squeezed out of the bond joint during the compression of the adhesive during bonding. A meniscus is formed at the chip periphery (see Fig. 4.20). This adhesive meniscus contributes as F_{circ} to F_{bond}. To maximize the bond strength, enough adhesive is needed to fill the gap between chip and substrate and to form a small meniscus at the chip periphery to have a F_{circ} contribution.

Very important is the similarity of mold bonding and UV-NIL. Excessive imprint resist is also squeezed to the mold periphery and forms a meniscus or an edge bead around the imprint field. This excessive material also contributes to the overall demolding force like F_{circ} (cf. Fig. 5.20).

4.3.3 Bond Joint Thickness and Mold Planarity

The mold planarity can be understood as the variation of the distance between the structured mold surface plane and the surface plane of the backside of the support. The smaller this variations is, the higher is the planarity of the bonded imprint mold. The higher the planarity is, the lower is the variation of the imprinted residual layer, which is one important criterion for high quality imprints.

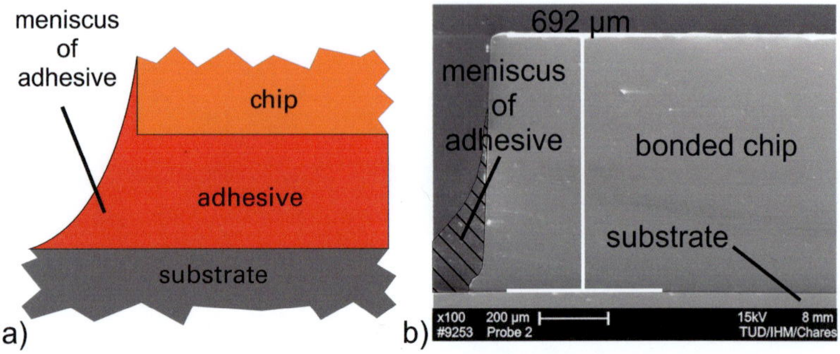

Figure 4.20: *a) Schematic overview of an adhesive BIM bond joint with meniscus and b) SEM of a cross section of a BIM with meniscus.*

Bond Joint Thickness The bond joint can be compared to the residual layer achieved with a flat and non-structured imprint mold. The bond joint thickness depended on the same parameters like the residual layer [cf. Eq. (3.3)]. In particular, this were the adhesive viscosity, the applied pressure during bonding, the adhesive volume (determines the initial adhesive thickness), the time which was used to compress the adhesive before curing and the size of the chip (determines the distance the adhesive has to flow to the chip edge) (cf. Sec. 3.3.4). Thus, thin bond joints can be achieved with low viscous adhesives, high bonding forces, a long time under high force before curing and low adhesive volumes. Additionally, planar bonding partners were the main requirements for thin and homogeneous bond joints. Bond joints with a thickness of about 0.4±0.2 µm were achieved with acrylate- and epoxy-based systems (see Fig. 4.21d and e).

Planarity Besides the homogeneous bond joint thickness, a high planarity and low thickness variation of both the chip and the support used were found to be crucial for the planarity of the final BIM. For example, a standard 150 mm fused silica wafer with a systematic total thickness variation of 10 µm diced into $10\times10\,\text{mm}^2$ small chips will lead to a thickness variation of roughly 0.7 µm per chip assuming a linear variation across the wafer. Other geometrical irregularities like the wafer bow or warp have to be accounted as well. According to the ITRS [1], a wafer-site flatness in the exposure field of less than 50 nm is acceptable after correction of linear failures. Thus, bow and warp effects in this $26\times10\,\text{mm}^2$ must be below 50 nm. In comparison to directly bonded molds, adhesives can partly level such surface irregularities. Further, the planarity of the bonding equipment was crucial.

The planarity of the final mold was directly connected to the bond joint thickness. On one hand, the bond joint thickness was important for the leveling of the surface waviness and the thickness variations of both bond partners. In principle, this enabled

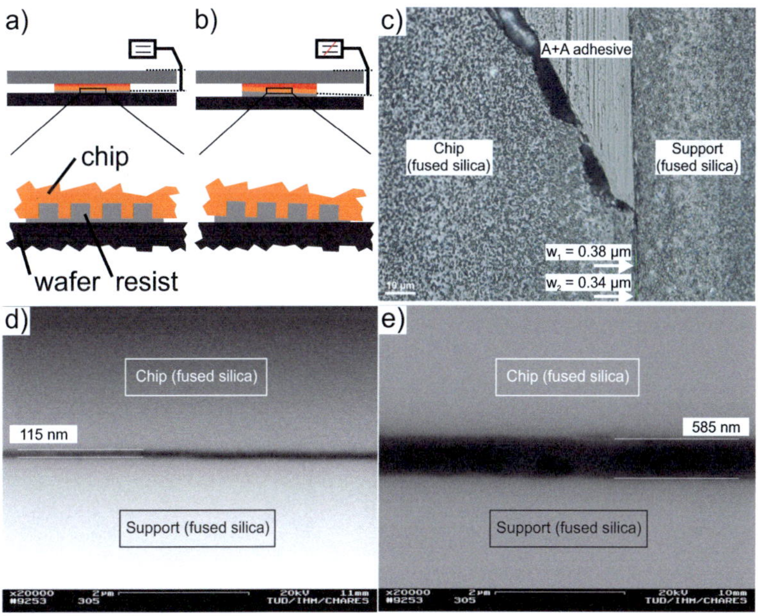

Figure 4.21: *Schematic overview of a BIM and the resulting imprint situation with a) a perfectly parallel BIM and a homogeneous residual resist layer and b) a BIM with non-planar chip and substrate surfaces and subsequently an inhomogeneous residual resist layer. Micrographs of adhesive joints made with c) an acrylate-bonded adhesive (visual light micrograph) and c-d) the same epoxy (scanning electron microscopy (SEM)) showing an exemplary difference of the thickness along the bond joint of 471 nm.*

the chip surface to be parallel to the support surface. Furthermore, thick bond joints can compensate more mechanical stress from differences in thermal expansion coefficients than thin bond joints. On the other hand, the thicker the bond joint, the higher was the probability for non.parallel bond joint and thus a planarity failure. A very thin bond joint was contrast to the bond strength, which increases with increasing adhesive volumes due to an increasing F_{circ}- contribution. An optimized adhesive volume was required to maximize the bond strength and to minimize the bond joint thickness. In practical bonding, this was achieved with a high-strength adhesive and a slight excess of adhesive at the periphery.

Laser profilometry and cross-sections gave evidences for influences of the mentioned geometrical irregularities of the bonding partners on the planarity. This planarity failures were independent of from the adhesive. For example, the difference between the maximum and the minimum of the surface plot was about 2 µm (see Fig. 4.22a). This represented the maximal deviation from an ideally planar surface. The maximal difference in the bond joint of the same sample was 0.5 µm (cf. Fig. 4.21d to e).

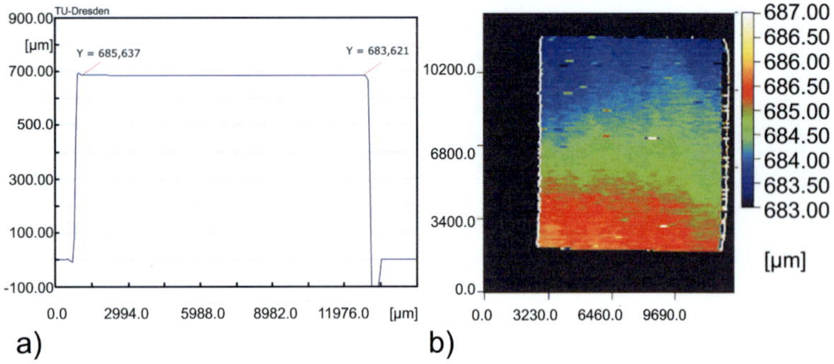

Figure 4.22: *Results of a scanning laser microscopy of an epoxy bonded mold with a) the profile of a line scan across the surface and b) the surface plot.*

Whereas the Fig. 4.22b represented a clear wedge defect, a typical dome-like height distribution was found for other BIMs (see Fig. 4.23). Possible explanations were a limited polymer flow from the center of the bond joint to the periphery inducing a bending of the chip. Additional bending could be induced by the cure-shrinkage of the adhesive. The difference between the highest and lowest point on the surface was 1.6µm (see Fig. 4.23a) for an average thickness of 1.7 µm (data obtained from cross-sections) as well as 1.5 µm (see Fig. 4.23b) for an average thickness of 1.9 µm (data obtained from cross-sections). For the latter sample a thickness difference of about 400 nm was found for the bond joint whereas a total difference between the highest and the lowest point on the surface was found to be about 900 nm. Thus, an additional influence of the chip and the support on the planarity independently from the adhesive must have been present.

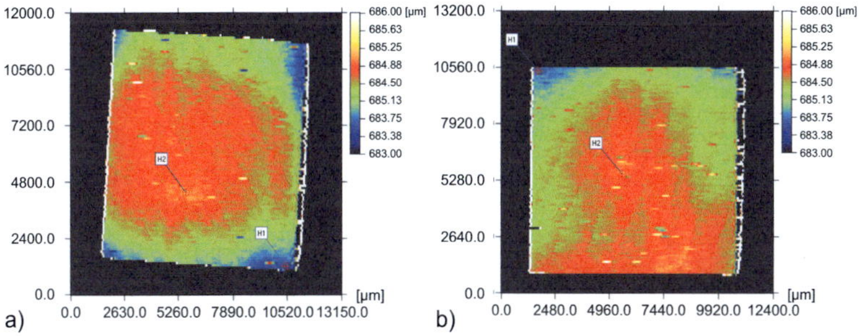

Figure 4.23: *Results of a scanning laser microscopy for two different samples bonded with Katiobond 4670 UV-curing epoxy: a) highest point H2: 685.1 µm, lowest point H1: 683.5 µm and b) highest point H2: 685.1 µm, lowest point H1: 683.6 µm. (modified from [140])*

The surface map in Figs. 4.22b and 4.23b could be not completely explained by a wedge error or a mold bow. For example the wedge error along x=3000 µm and x=10,000 µm in Fig. 4.22b is not identical. This was only explainable by an additional mold deformation, which could originate from small particles between the chip and the support but also from intrinsic geometric chip irregularities. A similar effect can be observed in Fig. 4.23b. By using well polished bonding partners, the planarity variations can be minimized. In summary, a maximum planarity error smaller than 1 µm should be achievable and thus enable BIM to be used for nanoimprinting. If an additional leveling like in the NPS300 of mold and substrate is possible, this would further compensate wedge errors of about 500 nm. However, non-linear variations of the surface planarity cannot be compensated. This problem does, however, also exist for monolithic imprint molds.

4.3.4 Chemical Stability

The BIM must withstand the following treatments and chemicals:

- mild removal particles and minor resist residues (e.g. by acetone),
- intensive removal of particles and resist contaminations (e.g. by acidic piranha, argon or oxygen plasma) and
- surface activation for antisticking layer coating (e.g. by acidic piranha or oxygen plasma).

The BIM must withstand these treatments to enable a large number of imprints without the need to be rebonded and thus a high throughput. Possible antisticking layer (ASL) degradation after only a few ten imprints is well known (cf. Secs. 2.9.3,4.1.3). A degraded ASL required a removal of the ASL remnants by a harsh cleaning and a ASL-recoating. Both the cleaning and the activation before the ASL-recoating are mainly done by oxygen plasma treatment. Acetone and acidic piranha was used to remove resist residues after imprinting. Acetone, acidic piranha treatments and oxygen reactive ion etching (O_2-RIE) was thus used to evaluate the chemical resistance of BIMs during imprint relevant treatments. Acetone dissolves adhesives or weakens the adhesive interface to the bonding partners due to swelling of the polymer matrix and replacement of physical bonds. O_2-RIE and acidic piranha gradually oxidizes polymers. Additionally, RIE mechanically sputters surfaces and removes minor residues or breaks bonds and thus eases a subsequent chemical attack.

Adhesive Stability Only the adhesive Vitralit was delaminated by a long acetone soaking (see Fig. 4.24a). The adhesive was not dissolved, which was visible due to adhesive residues on both bonding partners. Instead of that the adhesive failed cohesively after weakening by acetone. PDMS was probably also swollen leading to a reduced cohesive strength. O_2-RIE weakened the adhesive joints of two materials (A+A and Araldite) (see Fig. 4.24b). Compared to acidic piranha, the adhesive degradation by O_2-RIE was minor and led not to failed bond joints. The main reason for that was that the sputtering contribution of the O_2-RIE was mainly limited to the excess of adhesive on the chip periphery. Piranha cleaning was identified as the main challenge for multi-usable bonded molds. The A+A bonded BIM completely delaminated after only a few minutes in acidic piranha (see Fig. 4.24c). All other materials in Fig. 4.24 withstood the 10 min piranha cleaning. In general, the used acrylate based system (A+A) had the lowest resistance against piranha cleaning and epoxy based systems (EpoTek, Vitralit, Araldite) the highest. Often only the meniscus and a about 1 mm wide region under the chip starting from the chip edge was removed in piranha. Further prolonged piranha cleanings did not significantly increase the oxidation due to the limited diffusion of reactive species into and reaction byproducts from the thin adhesive joint gap.

The UV-curing Katiobond 4670 was introduced as alternative to the stable but thermally curing epoxies EpoTek and Araldite. Katiobond 4670 did neither show any evidences of reduced bonding strength upon prolonged acetone soaking nor during five subsequent piranha cleanings. Thus, Katiobond was identified and routinely used as BIM adhesive for further work. Typically, degradation started from the periphery and a continuous degradation and delamination of the bond joint could be observed for multiple piranha cleanings for less stable adhesives (see Fig. 4.25). Katiobond did not degrade to at such a high rate.

Increased Stability due to Fluorinated Silanes Fluorinated organosilanes were routinely used as ASLs (cf. Sec. 4.1). It was observed that vapor phase coated 1H,1H,2H,-2H-perfluorodecyltrichlorosilane (FDTS) significantly increased the chemical stability ob adhesively bonded imprint molds. An A+A adhesive joint withstood the 10 min piranha etch after vapor phase FDTS-coating. This was a tremendous increase in chemical resis-

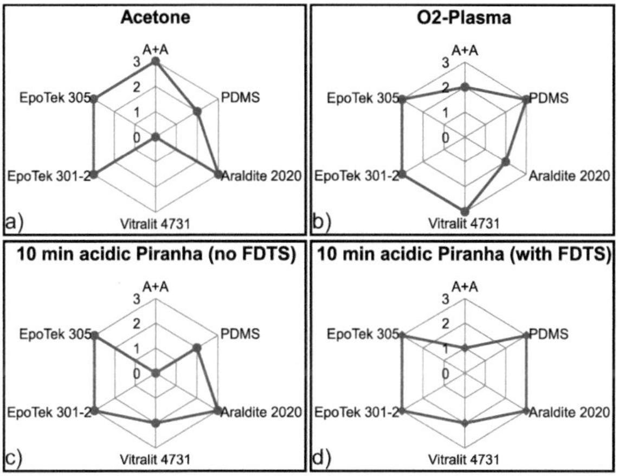

Figure 4.24: *Chemical resistance of selected adhesives from Fig. 4.19 against common treatments (0... not resistant, 3... no effect).*

tance. The protection effects of FDTS was even more obvious in a long term experiment. An FDTS coated BIM sample and a reference BIM sample bonded with the same adhesive (EpoTek 305) were treated with acidic piranha for multiple times. The piranha solution was refreshed every 10 min to maintain a high reactivity. The FDTS coated sample did no degrade during 6×10 min piranha treatment (Fig. 4.26a and b), whilst the uncoated reference samples continuously degraded (Fig. 4.26c and d). This was visible in an effective bond area and shear force reduction. The effective bond area was defined as ratio of the area of the adhesive not attacked during cleaning to the overall area of the bonded chip (cf Fig. 4.25).

The effective bond area of of uncoated BIMs was reduced by 17.2 % per piranha treatment (see Fig. 4.24a). The shear force was decreased by about 275 N per piranha treatment (see Fig. 4.24b). The observed plateau at about 1 kN in Fig. 4.26b was due to the measuring limit of the shear force sensor. The true values were expected to be much higher. Using the last four force and area measurements from Figs. 4.26a and b, the EpoTek305 adhesive had a high and for nanoimprinting more than sufficient average bonding strength with a 1σ-deviation of about $1.4\pm0.2\,\mathrm{kg/mm^2}$. However, this strength was only available for a quite limited amount of cleanings. In contrast, vapor coated FDTS significantly increased the chemical resistance of the EpoTek305 adhesive. A decrease was neither observed for the effective bonding area (see Fig. 4.26c) nor the shearing force for six subsequent piranha treatments (see Fig. 4.26d).

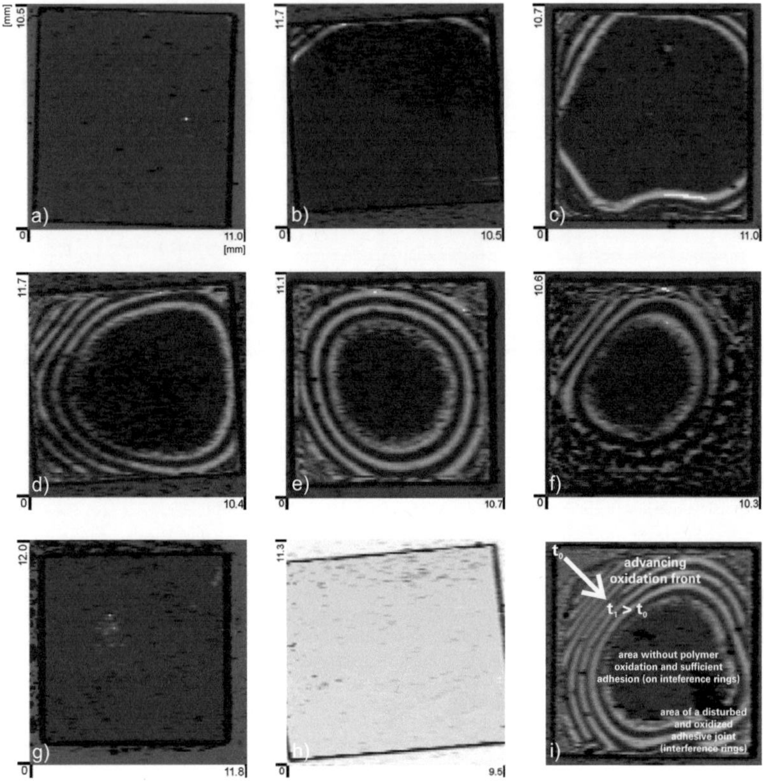

Figure 4.25: *Laser profilometry scans of piranha cleaning sequence for EpoTek305 with treatment times of: a) 1×10 min, b) 2×10 min, c) 3×10 min, d) 4×10 min, e) 5×10 min, f) 6×10 min. FDTS coated adhesives did not degrade during piranha cleanings. This was visible in comparison of an g) untreated sample and h) a sample treated for 6×10 min. Delaminated areas were visualized due to i) interference rings. (modified from [140])*

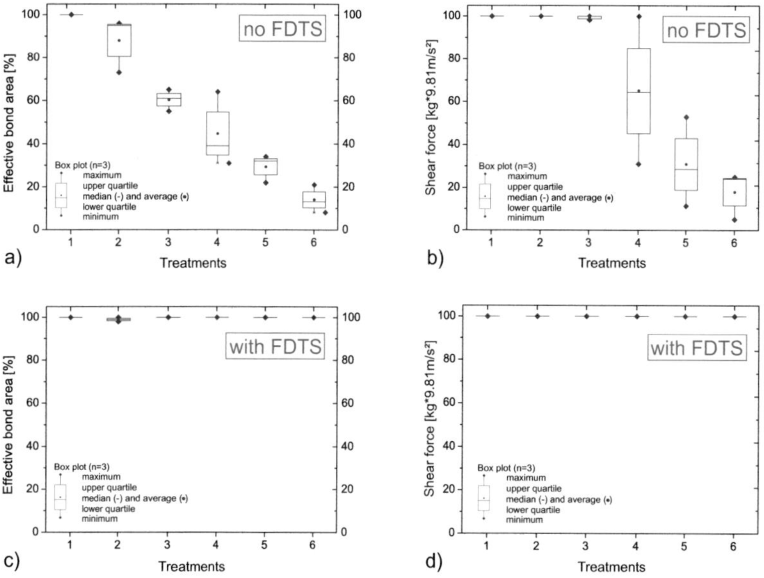

Figure 4.26: *Influence of FDTS on the chemical resistance of an EpoTek 305 based adhesive joint a-b) without FDTS and c-d) with FDTS (measurements correspond Fig. 4.25).*

Uncoated adhesive were quickly degraded at the periphery. Diffusion processes slowed the degradation rate in thin bond joints continuously down (see Fig. 4.27a). As already discussed, FDTS-layers were degraded by piranha treatment (cf. Sec. 4.1.3). The degradation rate depended on the coated material itself. As a consequence of the increased stability against piranha, a chemically stable FDTS-layer must have been formed on the adhesive (see Fig. 4.27b). The access to the piranha sensitive parts of the FDTS-chain must have been excluded or at least highly limited by densely packed FDTS-molecules and inferior wetting to form such a chemically stable layer. It was supposed that other fluorinated ASL show a similar behavior like the FDTS-ASL, if a densely packed layer is realized.

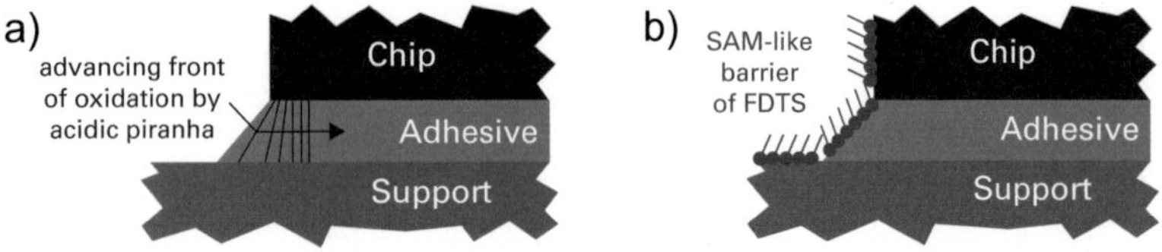

Figure 4.27: *Schematic of a) the progressive adhesive oxidation due to piranha and b) a model for the FDTS-barrier which limited the oxidation.*

Models for a Protective Layer Four different models were proposed for the FDTS-barrier layer. Firstly, the silane molecules were each bound to the adhesive via a Si-O-R binding, where R denotes the organic adhesive surface (see Fig. 4.28a). Second, the FDTS molecules were not linked to the adhesive at all and were instead hold in place by a 3-dimensional network of polymerized FDTS (see Fig. 4.28b). This 3-dimensional network spun the adhesive joint and was hold in place due to Si-O-Si bindings to the chip and the support as well as Si-O-Si bindings between FDTS-molecules. Third, the FDTS molecules were partly bound to the adhesive and formed a 3-dimensional, cross-linked network among each other (see Fig. 4.28c). Fourth, FDTS was not only bound to the surface but further migrated partly into the porous adhesive (see Fig. 4.28d), was bound there and also formed a 3-dimensional, cross-linked network.

It is known, that silanes can bind to activated polymer surfaces [118, 135]. However, bindings to carbon oxides are claimed to be less stable[135]. No special measures against cross-linking were made and at least some cross-links should have been present and additionally stabilized the FDTS-layer on the adhesive. FDTS-molecules cannot be cross-linked and surface bound at the same time[54]. The hydroxyl group density on the activated adhesive surface must have been comparable to silica to yield a dense and stable FDTS-layer like in model (a). This means, that model (a) is possible but as soon as cross-links were present, model (a) changes to model (c). Other models than (a) were favored. It is known that silanes do bind to polymers as mentioned above. The plasma treatment and activation before the vapor phase deposition enabled at least a partial binding of the FDTS-molecules to the adhesive. From these points of view, model (c) seems to be more probable than model (b). Assumed that the adhesive had large enough pores, model (c) can be extended to model (d). Thus, the latter two models seemed to be the most probable explanation for the observed protection behavior by FDTS. In

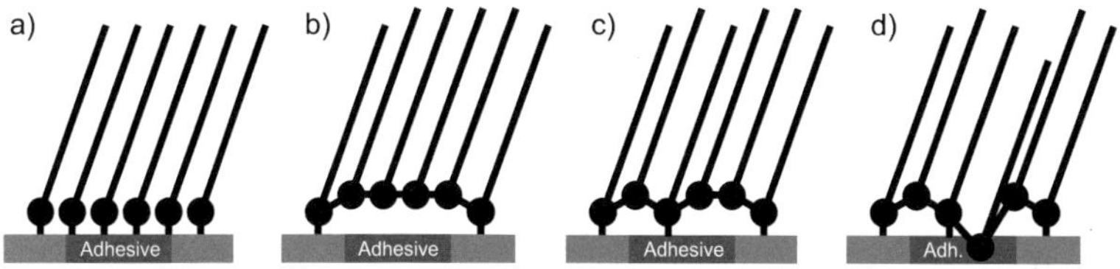

Figure 4.28: *Schematic representation of the proposed models for the FDTS-protection layer.*

summary, FDTS vapor coating enabled chemically stable but still demountable bonded imprint molds for multiple usages. To demount such molds, the FDTS had to be removed by RIE and the adhesive by piranha treatment. When the antisticking layer is refreshed, the protection layer on the adhesive is also refreshed. FDTS can prolong the lifetime for less chemically resistant adhesives.

4.3.5 Process Stability

Additionally to the chemical stability, the stability during imprint process was evaluated. The adhesive joint of a BIM must be stable during intensive UV-irradiation in UV-NIL and temperatures up to 250°C in T-NIL. To enable a long term use without rebonding, a high optical transparency and a high bond strength must be maintained throughout the mold lifetime. The optical transparency at 365 nm was determined via UV/VIS-transmission spectroscopy and the bond strength was determined with shear force measurements. The UV-induced aging was simulated with broad spectral, flood exposures under constant room temperature with doses ranging from 100 J/cm^2 up to 600 J/cm^2 (at 365 nm) simulating 100 imprints with 1 J/cm^2 to 6 J/cm^2 per imprint. Thermal aging was induced by heating samples to temperatures from 100 °C to 250 °C for 500 min to simulate 100 imprints for 5 min at the respective temperatures. This investigation was aimed at the thermal stability of the adhesive and neglected effects from multiple short T-NIL cycles.

The acrylate-based adhesives enabled very thin but also chemically not very resistant bond joints. The acrylates were useful for easily demountable BIMs. Two chemical resistant adhesives were compared for long-term use. The Katiobond 4670 was UV-curable and enabled a faster processing compared to the thermally curing of EpoTek305.

Thermally Curing EpoTek305 All samples were sufficiently transparent at 365 nm with more than 86 % transmission prior to the aging experiments. The initial bond strength was larger than 10 MPa. The EpoTek305 lap shear strenght was about 13 MPa and the die shear strength 23 MPa [141]. Thus, the experimentally obtained values fitted the datasheet values quite well. UV-aging had no influence on the transmission or bond strength of EpoTek305. In contrast, thermal aging revealed a long term stability against temperatures up to 160 °C for more than 500 min. Above 160 °C EpoTek305 lost its transparency due to blackening originating from adhesive carbonization. Thus, EpoTek305 degraded very fast at temperatures above 160 °C. The shear force was also reduced remarkably. Continuous operation is recommended below 100 °C [141]. For T-NIL applications above 160 °C, more resistant adhesives have to be used. However, EpoTek305 was well suited for UV-NIL applications.

UV-curing Katiobond 4670 A new material, Katiobond 4670 was found to be even more resistant than EpoTek 305. The shear strength is 35 MPa and the degradation is temperature 280 °C [142]. No UV-induced aging was found during typical imprints. Evidences for thermal degradation were visible for 250,°C but not for 200 °C. No reduction in the shear strength detectable for those temperatures but at 250 °C yellowing of the resist was observed. Thus, Katiobond was estimated as long term stable for T-NIL up to 200 °C.

4.3.6 Imprinting

BIMs were successfully used for imprinting of structures down to 350 nm line width over large areas with a 10×10 mm active imprint area (see Fig. 4.29). However, the resolution was limited by the available mold and no drawback was seen during frequent application of BIMs (cf. Sec. 4.34a,b or [143]).

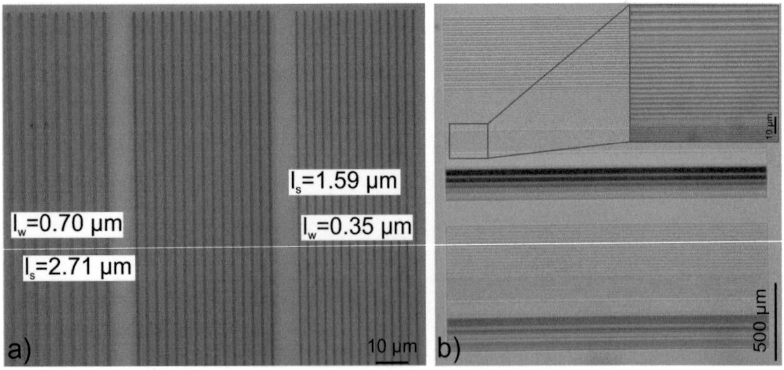

Figure 4.29: *Successful imprints with a) an acrylate-bonded mold (l_w … line width, l_s … line spacing) and b) an EpoTek305 bonded mold.*

In conclusion, BIM were routinely applied applied as promising alternative to monolithic molds throughout this work. BIMs are well suited for research, process development, product prototyping and even production if the appropriate chips, substrates and bonding equipment is used. Highly stable adhesives were beneficial when intensive cleaning was required after bonding.

4.4 UV-NIL Technique for Opaque Molds and Substrates

Basic Principle For state-of-the-art "industrial" UV-NIL and applied research, the required UV-transparent fused silica molds represent a major cost-driver (cf. Sec. 3.3.5). More cost efficient silicon molds would be desirable for UV-NIL. Non-transparent antisticking layers are an alternative to the predominant, degrading, fluorinated organosilane antisticking layers (cf. Sec. 2.7). To utilize opaque molds and antisticking layers, a solution different from the standard approach (see Fig. 4.30a) must be found to provide enough UV-intensity to the imprint material. One solution would be the usage of a transparent substrate [144]. Therefor, a transparent chuck and a UV-source below this chuck are required (see Fig. 4.30b). Another solution would be an inverted imprint setup, where the substrate is placed on top of the mold (see Fig. 4.30c). Unfortunately, both techniques are not very convenient for stamp-and-repeat imprinting (cf. Sec. 5.3). In usual UV-NIL stamp-and-repeat systems (through-mold-exposure), the UV-source is located above the mold and the substrate is located on a opaque, movable chuck (cf. Fig. 4.30a). An alternative technique developed in this work is non-transparent UV-NIL (NT-UV-NIL), which uses opaque molds (cf. Fig. 4.30d) in a standard through-mold-exposure imprint setup.

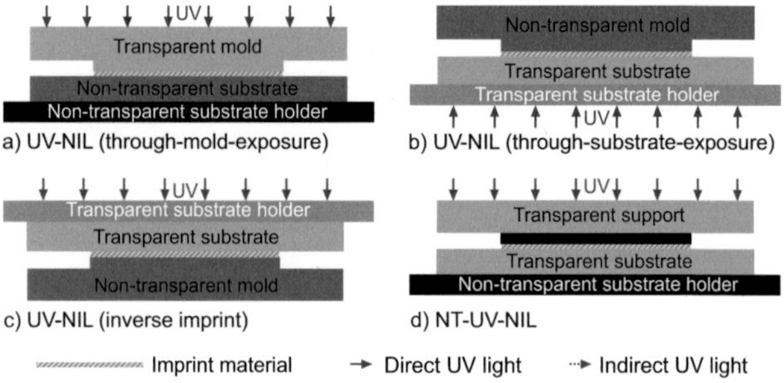

Figure 4.30: *Comparison of a) the standard UV-NIL setup and different principles for using opaque molds for UV-NIL: b) setup using an exposure unit located under the substrate, c) inverted setup with a mold being situated below the substrate and d) NT-UV-NIL using indirect exposure methods.*

The exposure occured not directly through the opaque mold but rather indirectly by diffuse light (see Fig. 4.31). To ensure sufficient curing, NT-UV-NIL uses small opaque pattern elements of, e.g., $10\times10\,\text{mm}^2$ area, bonded to a transparent support (cf. Sec. 4.3). Furthermore, transparent substrates or opaque substrates with at least a transparent layer in the micron range on top are required. Indirect exposure was realized in imprint setups offering a broad angular intensity distribution of UV-light. Such broad distributions were achieved by using a UV-source with a broad angular intensity distribution like a UV-LED-array (NPS300, SET S.A.S.) or an iron doped mercury arc lamp with light diffusive elements. Used diffusive elements were, e.g, an aluminum chamber, where a small laboratory press was placed in, or the wafer chucking (vacuum lines, holders, clamping) and interior of a standard contact mask aligner (flood exposure mode) (Al6-2, EV Group).

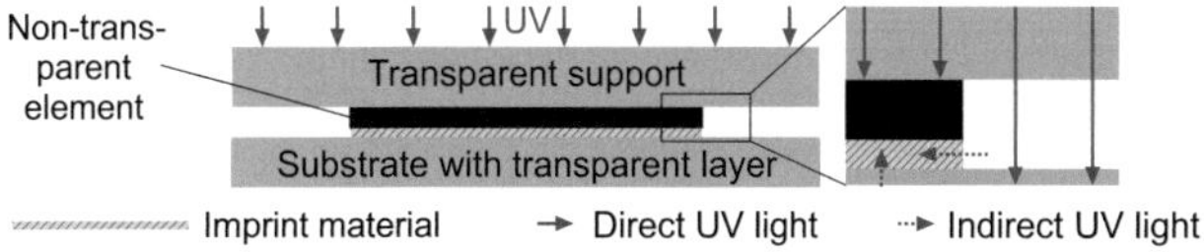

Figure 4.31: *Principle of indirect exposure (scattering, reflection, diffraction) exploited in NT-UV-NIL.*

Dose-Thickness-Relations The lowest curing dose for customized material was about 3 J/cm^2 (at 365 nm). Literature values for comparable monomer and photoinitiator systems were about 0.2-1.2 J/cm^2 [145–148]. The minimal dose for UVCur21SF was about 1.8 J/cm^2 (at 365 nm, see Sec. 5.2.2). For the customized material, below 30-35 µm thick transparent layers required about 20 times the minimal dose and at 9 µm thick SiO_2 layers already about 100 times the minimum dose was required (alway mercury arc lamp). The more sensitive UCCur21SF required at the same 9 µm SiO_2 only about 40 times the minimal dose (at 365 nm, see Fig. 4.32a). Complete curing could be not achieved with the sensitive UVCur21SF for a 4.5 µm SiO_2 layer (see see Fig. 4.32b). At least 9 µm SiO_2 layers were required. By comparison of a completely cured imprint field of 10×10 mm^2 opaque mold on an 18 µm SiO_2 layer (see Fig. 4.33a) with the same imprint on a 4.5 µm SiO_2 layer (see Fig. 4.33b) the lateral curing from the mold periphery to the mold center is clearly visible. As can be seen, mold features also influence the curing behavior. In conclusion, NT-UV-NIL requires a highly sensitive photoinitiator and transparent layers of at least about 20 µm for curing times being only a factor ten longer than standard UV-NIL[3].

Imprint Achievements and Applicability A 150 nm line was replicated from a mold feature larger than 400 nm in a 1 cm^2 silicon mold (see. Fig. 4.34a,b). Here, indications were found that features with a comparable size to the exposure wavelength might behave different than larger features. Micron sized waveguides were directly patterned and fully cured with NT-UV-NIL (see. Fig. 4.34c). Successful lift-off patterning was also demonstrated with NT-UV-NIL imprinted resists (see. Fig. 4.34d). Due to the required transparent layer, NT-UV-NIL is more suited for direct patterning and lift-off than for a pattern transfer to opaque substrates like silicon.

[3] An exposure intensity of 250 mW/cm^2 and an exposure time of 100 s was assumed.

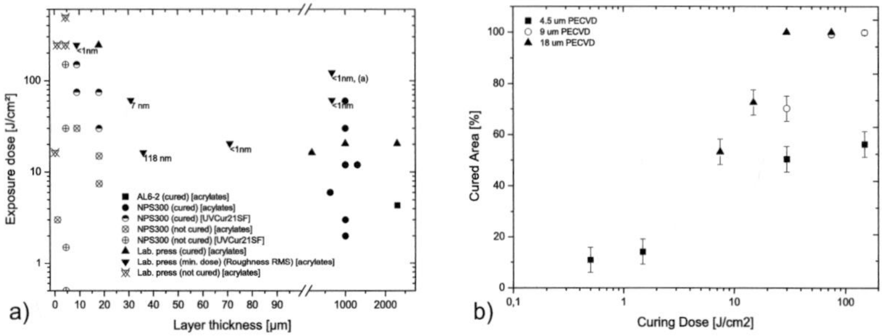

Figure 4.32: *a) Relation of the exposure dose to the transparent layer thickness for multiple experiments with different imprint setups and different transparent substrates and films. b) Imprint with the NPS300 system on a set of thin plasma enhanced chemical vapor deposited SiO_2 layers demonstrating the dependence of transparent film thickness, cured area and exposure dose. (mold size was always $10\times10\,mm^2$).*

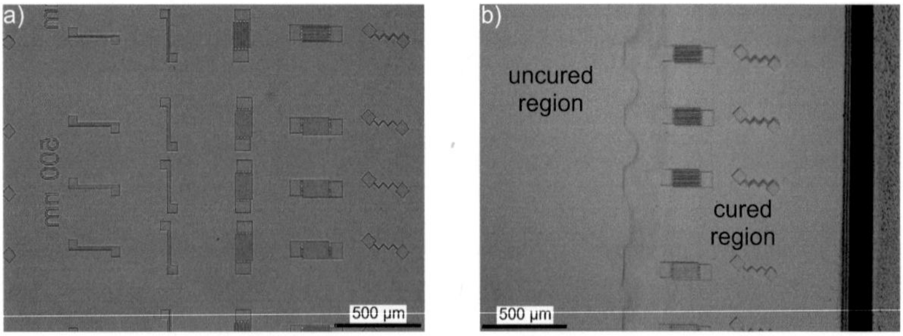

Figure 4.33: *a) Fully cured $1\,cm^2$ NT-UV-NIL imprint on an $18\,\mu m$ SiO_2 layer (mold center) and b) imprint with the same mold on a $4.5\,\mu m$ SiO_2 demonstrating the curing from the periphery to the center.*

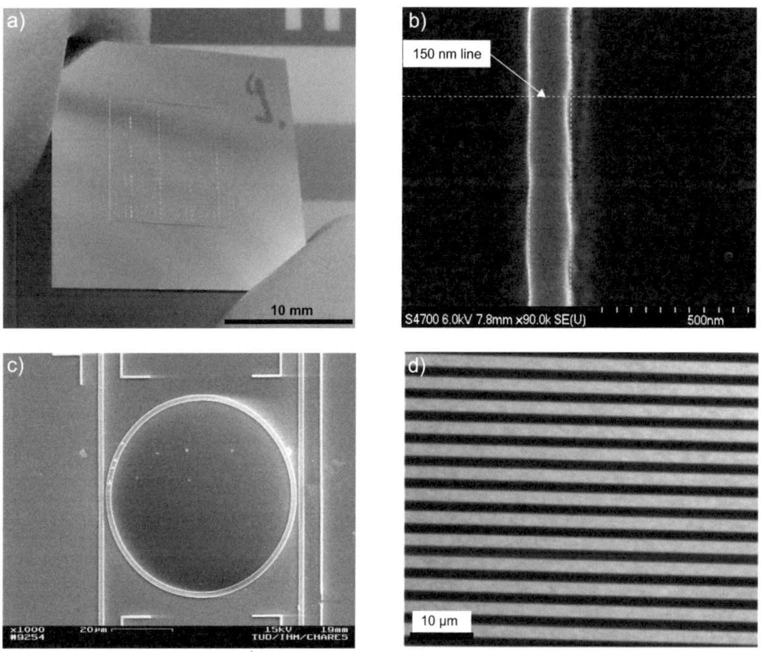

Figure 4.34: *a) Overview on an NT-UV-NIL imprint with a* $1\,cm^2$ *mold on an* $18\,\mu m$ SiO_2 *layer and b) a 150 nm line within this imprint field. c) MRR, directly patterned with NT-UV-NIL on a silica substrate (from [137]) and d) aluminum (bright) lines patterned by NT-UV-NIL and lift-off on a glass substrate.*

5 UV-NIL Technology for Polymer Microring Resonators

"In all science, error precedes the truth, and it is better it should go first than last."

(Hugh Walpole)

Different imprint techniques can be used to replicate polymeric microring resonators (MRR). T-NIL has been already demonstrated to yield high quality resonator structures for biological sensing [3, 17]. UV-NIL was favored in this work due to the ability to synthesize the required hydroxyl-functional imprint material for bioconjugation (cf. Sec. 4.2) and the simplification of the imprint process for highly negative molds (patterned area about 1.5 %) and micron-sized cavities.

Even though polymeric molds are currently gaining a huge interest, rigid molds were used in this work due to their compatibility with established resist systems (no oxygen cure-inhibition, no diffusion of resist components into the mold) as well as the ability to replicate MRR-structures with aspect ratios larger than five. Rigid molds are typically used in a stamp-and-repeat process, where a small reticle is imprinted multiple times on a larger substrate. The aim of this chapter was to develop and evaluate a stamp-and-repeat UV-NIL (SR-UV-NIL) technology for the MRR-fabrication with the NPS300 (SET S.A.S.) imprint system.

The chapter is structured in four parts. First, the material system is introduced. UVCur21SF was the main imprint material. Adhesion promotion by spin-coated, reactive trichlorosilanes was found to be too sensitive to environmental conditions. Reliable adhesion promotion was achieved with the commercial APS1 system. The second part comprises the pattern replication. Sub-300 nm MRR-gaps and aspect ratios larger than five were achieved. In part three, the developed SR-UV-NIL technology is presented. All imprints were performed without vacuum environment. Failures due to air trapping and insufficient cavity filling could be highly reduced for SR-UV-NIL in spin-coated films. The pre-exposure of subsequent imprint fields highly influenced the curing behavior in these fields. Pre-exposure can be reduced by optimizing the SR-UV-NIL imprint-field density. Local capillary filling of cavities was not observed. The major challenge for a successful technology remained the residual layer homogeneity. In the last part, residual layer free imprinting with hybrid imprint molds is discussed.

5.1 Material System

The MRR, as aimed for in this work, is based on single mode planar waveguides. The dimensions of these waveguides, their refractive index and the refractive index of the environment determine its guiding properties. The refractive index of typical substrates and UV-curing imprint materials is limited and restricts the possible waveguide dimensions. The MRR will be operated at a wavelength λ=1550 nm [21]. For a waveguide geometry of 2 µm width and 1 µm height, an imprint material with a refractive index at λ=1550 nm of about 1.56-1.58 is required for standard silicon oxide substrates ($n_{1550\,nm}$=1.444). For a low refractive index oxide ($n_{1550\,nm}$=1.400) an imprint material with a refractive index of about 1.51-1.54 is required [103, 149]. This will ensure both a guiding single-mode waveguide and a coupling micro resonator in a water environment operated at E_x- and E_y-polarisation.

Substrate Polymer waveguides on pure silicon are almost impossible due to the high refractive index of silicon of 3.48 being well above typical values of polymer materials. A low refractive index undercladding or substrate was required to allow for a broad spectrum of waveguide materials and dimensions. A typically low refractive substrate is fused silica as an oxide with one of the lowest refractive index (see Tab. 5.1). To reduce the refractive index of any material, fluorine can be integrated [16]. This is a general approach used for metals, resulting in compounds like MgF_2, or polymers (cf. Tab. 5.1). MgF_2-substrates are quite expensive and rarely available in wafer sizes of 6 in as would be required for the MRR-production. Fluorinated polymers can be applied as undercladding for the MRR but require the incorporation of adhesion promoting groups. Fluorination of polymers leads in general to a low SFE (cf. Tab. 2.4) and therefore to a low adhesion of imprint materials to the fluorinated undercladdings.

Table 5.1: *Refractive indices of exemplary substrates and under-claddings being available for MRRs.*

Material	n (λ=1550 nm)	Reference
Porous, nanorod based SiO_2	1.054**	[150]
Low-k spin-on dielectrics	1.17-1.39***	[151]
Fluorinated polymers	1.375-1.418*,**	[152]
MgF_2	1.371	[153]a
Fused silica	1.444	[153]b
Si_3N_4	1.973	[153]c
Si	3.476	[153]d

* monomer properties, ** refractive index at n_D=589 nm (at 1550 nm even lower)
*** refractive index at 633 nm (at 1550 nm even lower)

Thus, fused silica or at least silicon wafers with about 10 µm silicon oxide on top as under-cladding were required for low loss waveguides. To further enhance the light confinement and to decrease the substrate leakage of waveguides on silicon oxide, a wet chemical under-etch of the silicon oxide under-cladding can be used [17].

Imprint Material Typical polymers and imprint resists have refractive indices of 1.4-1.6 (see Tab. 5.2). High refractive indices can be found for the T-NIL materials polystyrene (PS) and polycarbonate (PC). The highest refractive index of UV-NIL materials was found for a polyimide and OrmoCore (micro resist technology GmbH), a commercial, organic-inorganic hybrid material designed for optical waveguide manufacturing by UV-lithography. Another very interesting material would be the UV-NIL material TPIR-305 (Tokyo Ohka Kogyo Co.) with a refractive index of n_{633nm}=1.6 being much higher than that of OrmoCore. Unfortunately, the material structure is not suited for a wavelength of 1550 nm[1]. The next highest refractive index was found for the commercial mr-UVCur06 (micro resist technology GmbH). It has a relatively high refractive index increase upon curing compared to, e.g., OrmoCore. However, UVCur06 is not recommended for high aspect ratio features [154]. The commercial material mr-UVCur21SF (micro resist technology GmbH) has a medium refractive index compared to the other materials. OrmoCore has a relatively high viscosity of 3.1-3.9 Pa·s and is not applicable for ink-jet dispensing. UVCur21SF instead was found to be ink-jet dispensable and can be spin-coated. The customized formulation A+A [(tert-butyl acrylate)-co-(ethylene glycol diacrylate)] (cf. Fig. 4.7) established in this work had a comparable low refractive index.

Additional efforts were made to increase the refractive index of the customized acrylate-based material by co-polymerization of styrene monomers as a high refractive component with acrylate or methacrylate monomers as quick UV-curing component. The theoretical refractive index was calculated from the monomer refractive indices using the Lorentz-Lorenz formalism for homogeneous mixing [see Eq. (5.1)] and an assumed increase of the refractive index upon curing of 1-2 %.

$$\frac{n_{12}^2 - 1}{n_{12}^2 + 2} = \epsilon_1 \frac{n_1^2 - 1}{n_1^2 + 2} + \epsilon_2 \frac{n_2^2 - 1}{n_2^2 + 2} \tag{5.1}$$

n_{12} is the effective refractive index of the mixture of the monomers having each a refractive index n_1 and n_2. ϵ_1 and ϵ_2 are the respective volume fractions of the monomers in the mixture. Up to 60 Vol.% styrene monomer were successfully co-polymerized with ethylene glycol dimethacrylate (EGDMA) as a cross-linking monomer. However, only less than 40 Vol.% styrene could be successfully co-polymerized with tert-butyl methacrylate (tBMA) which forms a linear, non-cross-linked co-polymer. Further investigations on the co-polymer structure and the realized refractive index of the resulting material are required but were beyond the scope of this work. However, this is an interesting method to modify the refractive index of the imprint material and to find a UV-curing, thermoplastic material which could be smoothed in an annealing step after imprinting, comparable to T-NIL patterned MRRs [17].

[1] According to information of the Tokyo Ohka Kogyo Co. R&D-department

Table 5.2: *Refractive indices of typical NIL-materials at a wavelength of λ=1550 nm.*

#	Material	NIL-technique	n (λ=1550 nm)	Parameters / Ref.	Comment
1	Polyimide	UV-NIL	1.583-1.608	[155]	-
2	SU-8	T-NIL	1.570	[156]	NANO SU-8-25
3	Polystyrene (PS)	T-NIL	1.555	[153]e	n extrapolated
4	Polycarbonate (PC)	T-NIL	1.544	[153]e	n extrapolated
5	OrmoCore (cured)	UV-NIL	1.536	[157]	n extrapolated
6	Styrene-co-EGDMA (cured)	UV-NIL	1.528-1.543*	62.6 Vol.% St. 35.4 Vol.% EGDMA 2.0 Vol.% PI	minimal defects
7	OrmoCore (uncured)	UV-NIL	1.518	[157]	1.17 % increase upon curing**
8	mr-UVCur06 (cured)	UV-NIL	1.518	[158]	n extrapolated, 1.24 % increase**
9	mr-I T85	T-NIL	1.517	[159]	n extrapolated
10	Polyimide	T-NIL	1.513-1.546	[160]	T_g>330 °C
11	Styrene-co-EGDMA (cured)	UV-NIL	1.509-1.523*	41.4 Vol.% St. 55.8 Vol.% EGDMA 2.8 Vol.% PI	sucessfully polymerized
12	Styrene-co-tBMA (cured)	UV-NIL	1.506-1.521*	57.4 Vol% St. 40.6 Vol% tBMA 2.0 Vol.% PI	not completely cured
13	mr-UVCur21SF (cured)	UV-NIL	1.491-1.492	[161]	1.17-1.24 % increase assumed***
14	Styrene-co-tBMA (cured)	UV-NIL	1.482-1.496*	39.2 Vol% St. 58.9 Vol% tBMA 1.9 Vol.% PI	relatively soft material
15	mr-UVCur06 (uncured)	UV-NIL	1.481	[162]	n extrapolated
16	mr-UVCur21SF (uncured)	UV-NIL	1.474	[161]	n extrapolated
17	Polyacrylate (PA)	T-NIL/UV-NIL	1.467	[153]e	n extrapolated
18	Polymethylmethacrylat (PMMA)	T-NIL	1.461	[153]e	n extrapolated
19	Styrene-co-tBMA (cured)	UV-NIL	1.456-1.470*	18.8 Vol% St. 78.8 Vol% tBMA 2.4 Vol.% PI	sucessfully polymerized
20	Customized resist A+A (Polyacrylate)	UV-NIL	1.453-1.456	-	n extrapolated

* estimated from monomer properties and 1-2 % increase upon curing, ** from extrapolated data from the mentioned references , *** based on #7,8
St.-Styrene monomer, tBMA-tert-butyl methacrylate monomer, EGDMA-ethylene glycol dimethacrylate monomer
PI-photoinitiator 2-hydroxy-2-methyl-propiophenone

In conclusion, no UV-NIL material was found to meet the requirements on standard fused silica surfaces for the given layout except one polyimide[2]. The layout was specified for a minimal critical dimension of 200 nm, which limited in turn the possible etch depth in fused silica to 1 µm and resulted in $2\times1\,\mu m^2$ non-quadratic waveguides. Increasing the etch depth relaxes the required refractive index. A 2 µm quadratic single-mode waveguide would be feasible on standard silicon oxide as substrate for an imprint material with a refractive index of 1.51-1.54 [103, 149]. Achieving a high etching depth for small critical features in silicon is less challenging than in fused silica. Thus, a UV-NIL process using silicon molds would be much more straight-forward (cf. Sec. 4.4).

Adhesion Promotion Sufficient adhesion of the resist was assured by two different adhesion promoters. The reactive silane TCSP-MA was used as 1 mM spin-on solution in isooctane giving very thin layers, but not necessarily monolayers, on O_2-plasma activated surfaces. A disadvantage of TCSP-MA was its sensitivity against environmental moisture leading to a precipitation of the solution in about 1-2 weeks under standard clean-room conditions. This reduced the effective silane concentration and generated particulates on the substrate surface which ultimately led to adhesion failures and imprint defects. As an alternative, APS1 (micro resist technology GmbH) was used. This spin-on solution resulted in about 13 nm thick layers at 5000 rpm and exhibited excellent adhesion promotion properties for several polymer systems (UVCurSF21, NIF5a, A+A, epoxy adhesive). The environmental stability, handling and processing of APS1 was found to be much better than for TCSP-MA. APS1 was the standard adhesion promoter during this work. APS1 provided a stable coupling of UVCur21SF waveguides to the substrate even during complete immersion in water.

[2]Material was not available when experiments were performed.

5.2 MRR Replication

The used imprint system NPS300 (S.E.T SAS) was designed for monolithic molds having a size of 65×65 mm^2 and a thickness of 6.35 mm (see Fig. 5.1). The standard manufacturing process of the mold vendor (Photronics MZD and Jenoptik Laser, Optik, Systeme GmbH) comprised the parallel processing of four molds on one 6 in photomask blank and a subsequent dicing into four molds.

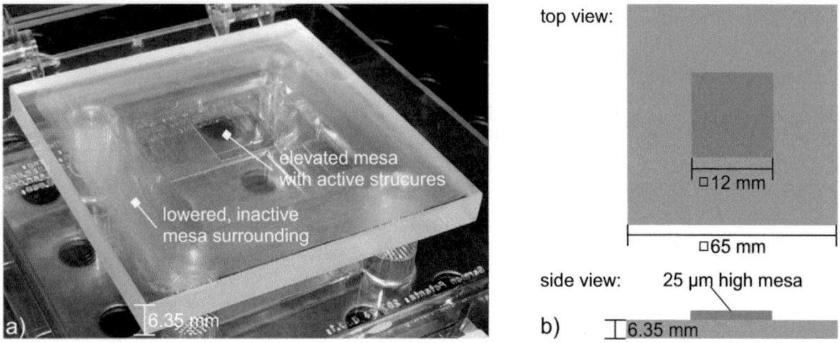

Figure 5.1: *a) Photograph of an exemplary monolithic mold in a storage box and b) dimensions of the monolithic mold.*

5.2.1 Mold Manufacturing

Different MRR test structures were put into a 12×12 mm^2 mold layout. The layout was directly electron beam written by the vendor into a ZEP7000 electron beam resist on a 70 nm thick antireflective chromium on a fused silica substrate (Hoya Corp.). The resist pattern was transferred into the chromium by dry etching. The chromium was then used as hard mask for a 1 µm deep dry etching into the fused silica. Subsequently, the hard mask was removed from standard molds and left on HNMs (cf. Sec. 3.3.3). The 12×12 mm^2 wide mesa was centered on the monolithic mold and produced by lowering the area around the mesa by a dedicated mechanical milling. During mesa formation, the mesa itself was protected from particles by a glass cover. The designed had a structured area ratio[3] of about 1.5 % in the active imprint area (see Fig. 5.2). Molds with an area ratio below 50 % are termed negative molds in this work.

The final fused silica mold provided resolved 200 nm MRR-gaps (see Fig. 5.3b). For the chromium hard mask, it was assumed that 1/3 of the antireflective layer was comprised of an oxynitride layer and 2/3 was metallic chromium. The oxynitride was completely removed during the oxide etch of fused silica leaving about 55 nm metallic chromium,

[3]Ratio of recessed to elevated area on the mold in %.

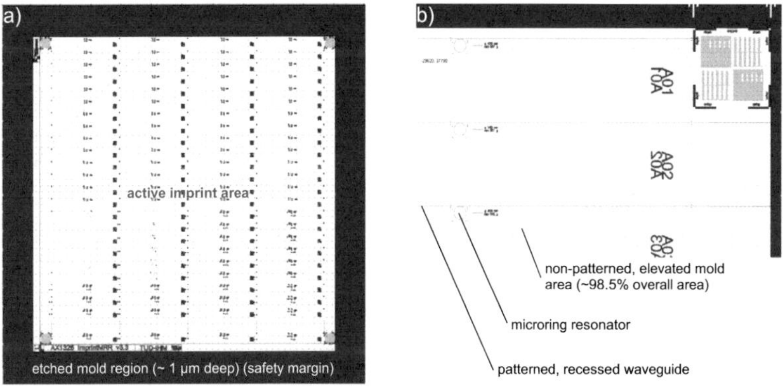

Figure 5.2: *a) Overview on an MRR-mold layout with the 1 mm wide safety margin (etched together with active imprint area) and the 10×10 mm² active imprint area. b) Close-up of mold layout emphasizing the highly negative mold pattern (1.5 % patterned area).*

which was determined by differential profilometry of molds with and without chromium.

5.2.2 Pattern Transfer to Imprint Material

The minimal curing dose for UVCur21SF is 0.7 J/cm² in vacuum [161]. Under ambient air conditions, the minimal curing dose was found to be about 1.8 J/cm². For a reliable curing, 2.0 J/cm² (14 s at 140 mW/cm²) was used as standard exposure dose. The increased minimal curing dose compared to vacuum resulted from oxygen inhibition and was consistent with other published data on comparable resist systems (cf. Sec. 4.4 for detailed discussions).

The typical waveguide after replication was characterized by three distinct features (see Fig. 5.4):

- footing on the interface to the substrate (origin: mold),
- edge super-elevation on the waveguide top (origin: mold and shrinkage),
- concave bow of the waveguide sidewalls and the waveguide top (origin: shrinkage).

The waveguides were designed as 2 µm wide channels in the mold. A SEM inspection of the mold (cf. Fig. 5.3b) yielded a different waveguide width at the top and the bottom of the channel (see Tab. 5.3). Mold (see Tab. 5.3) and imprint (see Fig. 5.4, Tab. 5.4) were compared for the replication accuracy: The bottom width of the mold

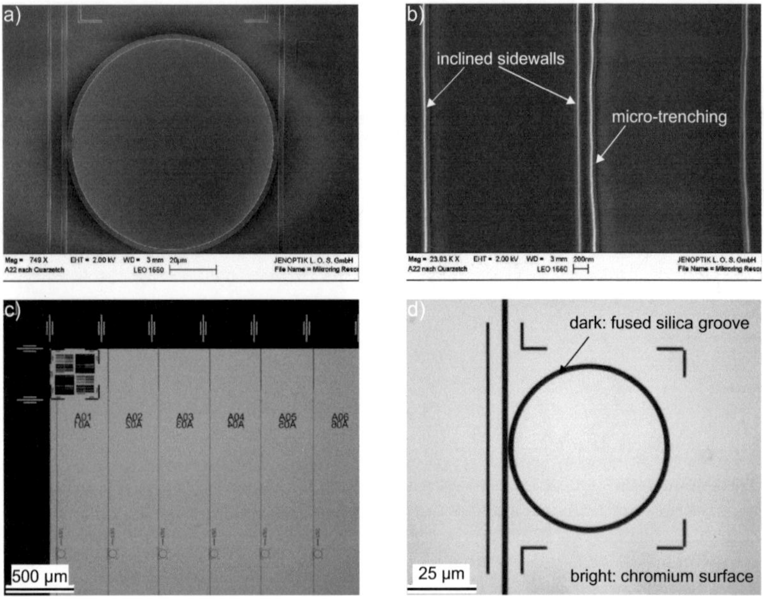

Figure 5.3: *a-b) Scanning electron micrograph of the final monolithic mold with removed chromium hard mask and an MRR with a 200 nm gap and c-d) optical micrographs with an overview and details of a 200 nm gap MRR with the chromium hard mask still on top to be used as HNM (bright areas = chromium, dark areas = fused silica).*

(1941±16 nm) was comparable with the top width of the waveguide after imprinting (1850±51) nm. According to these averaged values, a lateral shrinkage of 5 % was present. The depth of the mold (984 nm[4]) was well comparable with the central waveguide height after imprinting (982±25 nm). No significant shrinkage could be observed here. The difference between top and bottom waveguide width in the mold was attributed to a sidewall inclination. The sidewall angle of 86 °determined from the imprint (see Fig. 5.4) explained well the observed difference between top and bottom of about 70 nm. The footing of the waveguide after imprinting was quite large and could not be explained with the dimensions of the waveguides found on the mold[5]. Additionally, the shrinkage was found to be at moderate 5 % and could not explain this footing. Most probably, the footing was already present in the mold even though it was not obvious in the top-view scanning electron microscopy (SEM) inspection (cf. Fig. 5.3).

Table 5.3: *Mold characteristics determined by profilometry and SEM measurements.*

Parameter	AVG [nm]	$\pm\sigma$ [nm]
Waveguide depth (stylus profilometer)	984	-
Waveguide width at mold bottom (top-view SEM)	1941	16
Waveguide width at mold top (top-view SEM)	2082	23
Difference top-bottom p. edge (top-view SEM)	70	-

The edge-superelevation (see Fig. 5.4) might have its origin in a microtrenching at the mold bottom during fused silica etching. However, the continuous slope of this edge-superelevation emphasizes an additional component like shrinkage resulting in an inwards bow of the surface. Similarly, the slight inwards bow of the sidewall was also expected to be a direct result of the resist shrinkage. A discontinuity, a slight step, of the imprinted sidewall profile was observed (cf Fig. 5.4a) and attributed to the mold etch process. Indeed, such steps were found for silicon etching processes which were interrupted for etch depth control.

The smallest imprinted MRR gap was about 280 nm, which corresponded to an aspect ratio of Ω=3.5 (see Fig. 5.5a). The gap in the mold was about 200 nm indicating a lateral shrinkage of the waveguide of about 4 %. This was consitent with the theoretical lateral shrinkage of, e.g., butyl acrylate in the order of 5 % [163]. The smallest gap of 100 nm on the mold was not resolved due to sidewall inclination during fused silica etching. However, close to the defect a gap of about 100 nm was resolved (see Fig. 5.5b). This corresponded to an aspect ratio of Ω=9.8. Due to the resist shrinkage, the demolding of such high aspect ratio gaps between micron-sized structures does not pose a problem for UV-NIL. Thus, MRRs with aspect ratios of ten can be achieved as long as the gap is resolved in the mold.

The demolding of high aspect ratios was less difficult for convex mold structures than for concave mold structure (= waveguides). Waveguide replication with moderate aspect ratios of about Ω=2 was demonstrated (see Fig. 5.6). Thus, quadratic waveguides of 2 µm

[4]provided by the mold manufacturer

[5]SEM micrographs were provided by the mold manufacturer.

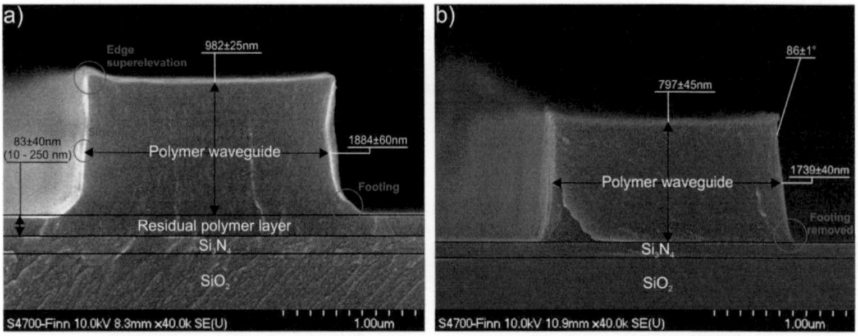

Figure 5.4: *Cross sections of a polymer waveguide with typical dimensions and their statistical variations: a) directly after imprinting and b) after O_2 reactive ion etching used to remove the residual layer.*

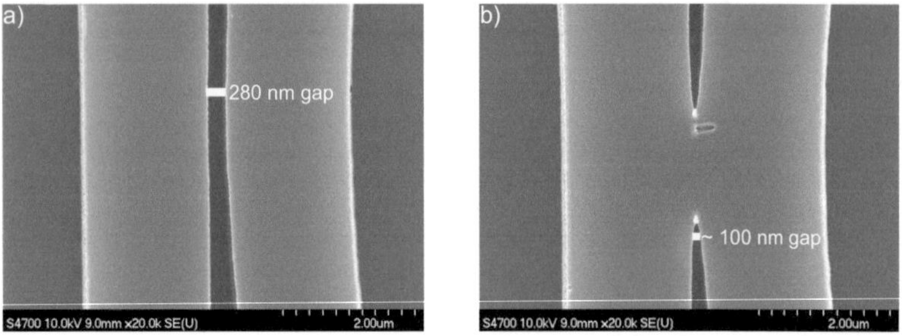

Figure 5.5: *a) Smallest resolved MRR-gap of 280 nm and b) smallest gap found close to a defective 100 nm (design) gap originating from mold etching.*

width and height are feasible. Smaller waveguides are not required due to the refractive indices of available polymers and substrates.

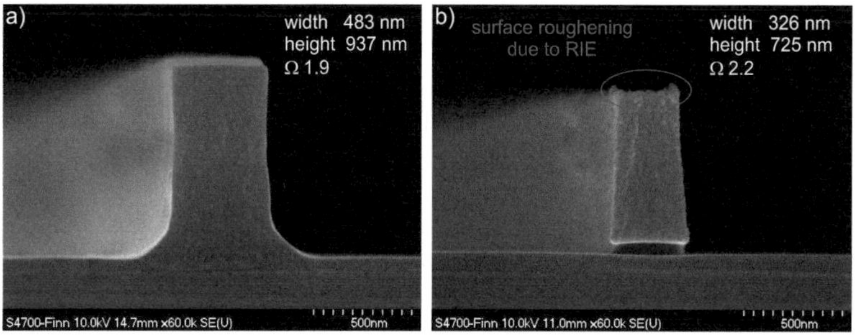

Figure 5.6: *Small polymeric waveguide: a) post-imprinted / before plasma treatment and b) after plasma treatment. Note the roughening of the top surface due to plasma treatment.*

Highly negative imprint molds were found to be quite resistant against mechanical defects in the mold. Especially during material and imprint parameter evaluation the mechanical impact on the mold is comparably large. Defects led to a mold deformation but only in very rare cases to mold damage. Also intensive mold cleaning processes never led to structure defects. In contrast, mold areas with positive character were prone to serious mechanical defects after short term use. Especially during mold cleaning, small isolated features of aspect ratios around two (500 nm feature size) were broken off very often. In contrast, small MRR-gaps were found to be rather mechanically stable because the gap was not isolated and instead stabilized by a large, surrounding unpatterned area. The fact of isolation was the crucial point.

5.2.3 Residual Layer Removal: Effects on Waveguides

The imprint inherent residual layer is normally removed by an anisotropic reactive ion etching (RIE) process. Parameters of the standard O_2-RIE used throughout this work were 60 sccm O_2 gas flow, 10 Pa (75 mTorr) chamber pressure and 67 W HF-power resulting in 350-400 V self-bias (Oxford Plasmalab 80 PLUS). The etch process had an anisotropy factor, defined as vertical etch removal to lateral etch removal, was 1.7 (see Tab. 5.4). The etch rate was about 100 nm in 22 s. After O_2-RIE, the edge-superelevation and the inwards bow was still present. The residual layer and the waveguide footing (see Fig. 5.4b and Tab. 5.5) were removed completely and to a great extent, respectively. An almost perfectly shaped waveguides were obtained. However, due to the anisotropic etching, the waveguide width was also reduced. The high residual layer thickness and variation in Fig. 5.4 originated from the mold and will be discussed later on.

The etch removal obtained from SEM and profilometry were quite comparable. The

Table 5.4: *Typical geometrical dimensions of polymer waveguides directly after imprinting and after a plasma residual layer etch determined from about 20 SEM cross-sections. (RLT: residual layer thickness)*

		post UV-NIL		post O_2-RIE	
		AVG	$\pm\sigma$	AVG	$\pm\sigma$
RLT [nm]		83	40	0	0
	top	1850	51	1701	40
WG width [nm]	center	1884	60	1739	40
	bottom	2358	75	1851	49
	left	1033	36	847	50
WG height [nm]	center	982	25	797	45
	right	1010	30	837	46

measured deviations included also systematic measurement errors, which were typically larger for profilometry. Both SEM and profilometry suffered from the drawback, that a measurement could not be repeated exactly at the same position before and after RIE due to a modification of the region of interest during the measurement.

Table 5.5: *Parameters derived from multiple SEM cross-sections to characterize the impact of the standard O_2-RIE residual removal process on polymer waveguides (etch time 3×22 s). (RLT: residual layer thickness)*

Parameter	AVG [nm]	$\pm\sigma$ [nm]
Etch removal vertical (SEM) (2×22 s)*	183	54
Etch removal vertical (Stylus profilometer) (2×22 s)*	178	106
Etch removal vertical r_o (SEM)	183	54
Etch removal lateral r_l (SEM)	153	70
Footing per edge (post imprint)	237	-
Footing per edge (post O_2-RIE)	65	-
Edge super-elevation (post imprint)	39	-
Edge super-elevation (post O_2-RIE)	44	-

* The value equals etching for about 2×22 s because in the first 22 s only the residual layer was etched.

The used RIE-process highly increased the surface roughness to a peak-to-peak value of about 30 nm (see Fig. 5.7). Even though UVCur21SF is a proprietary resist formulation, a co-polymer is highly probable, which might have led to different etch rates of the different co-polymer components at the nano-scale and ultimately to the rough, granular surfaces.

Depending on the waveguide design, the sidewall roughness, especially for resonator structures and bended waveguides, is more loss-relevant than the top and bottom waveguide roughness [17]. Thus, the roughness increase was not seen as a limiting factor for the waveguide production. However, optimized etching processes and thinner residual layers are required for a reduced surface roughness enabling reduced waveguide losses.

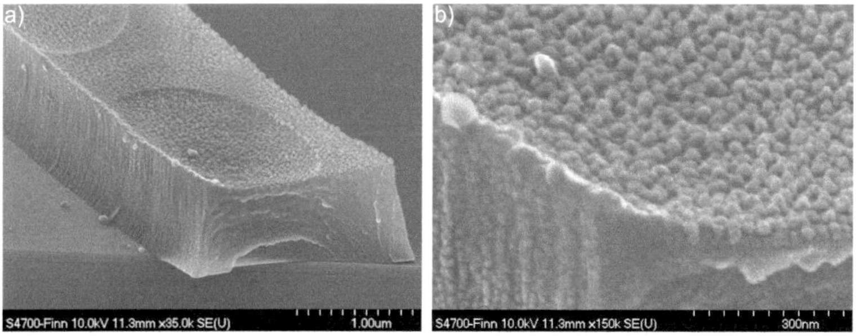

Figure 5.7: *a-b) Surface roughening (30 nm peak-to-peak) due to O_2-RIE residual layer etching of UVCur21SF polymer waveguides. Compare Fig. 5.5 for the surface roughness before plasma and additionally Fig. 5.6 for the roughness increase. (Note that the substrate in a) showed a much smaller grain size. The roughness was, thus, not due to the gold sputtering for SEM measurements.)*

5.3 Stamp-and-Repeat UV-NIL

After demonstrating the ability to replicate the required MRR-structures in single imprints, the feasibility of a medium-volume MRR-production had to be evaluated. Stamp-and-repeat nanoimprint lithography (SR-NIL) is a replication technique for rigid imprint molds to ensure a high level of planarity and a homogeneous residual layer thickness (RLT). In contrast to wafer-scale imprinting, SR-NIL repeats an imprint with the same mold step-by-step over a substrate. Often drop-on-demand dispensing (DOD) is used to apply an optimized resist volume and pattern. DOD requires an intensive, layout dependent process optimization for air bubble free imprints [104]. Spin-coating of low-viscous resists as homogeneous, wet films followed by a SR-UV-NIL process [164] reduces the effort to optimize the initial resist layer. Dispensing and spin-coating are compared in Sec. 5.3.1. New challenges arose for SR-UV-NIL in spin-coated films and will be discussed in the following sections. Structure filling was investigated for spin-coated films and revealed that thicker initial films improved the cavity filling (see Sec. 5.3.2). Most challenging was the pre-exposure of adjacent imprint fields (see Sec. 5.3.3). Neighboring exposure effects led to an adhesion promoter degradation and a different curing behavior of the resist. Additionally a large RLT-inhomogeneity was found for the used spin-coated films (see Sec. 5.3.4).

5.3.1 UV-NIL Defects

The yield and typical defects for repeated imprints were used to evaluate optimal resist application methods for the MRR-fabrication. To characterize the imprint failures, a qualitative failure description was established (see Tab. 5.6). The residual layer homogeneity is separately discussed in Sec. 5.3.4. Curing defects are mainly discussed in Sec. 5.3.3. In the following structure defects are discussed mored detailed.

Observed Structure Defects Typical defects comprised missing parts of replicated features to different degrees originating from two failure modes. In closed cavities like ring waveguides, often a single defect located at one waveguide sidewall was found (see. Fig. 5.8b) originating from bubble trapping. The second failure mode was self-assembling and incomplete cavity filling. Typical failures comprised deep (see Fig. 5.8c) or shallow (see Fig. 5.7a, 5.8f) defects being centered along the waveguide. Instabilities of the low-viscous imprint material and an incomplete cavity filling might have lead to self-assembling structures within the waveguide channels. Incomplete filling was also the reason for inhomogeneous gap regions (see Fig. 5.8d). Also rough and wavy sidewalls were found (see Fig. 5.8e,f) originating probably from self-assembling along the sidewall and an insufficient polymer-flow into the cavities. Adhesion failures were quite seldom. Often the structures were only lifted from the substrate but not completely ripped indicating well tuned adhesion promotion and antisticking measures for most imprints (cf. Fig. 5.8g through i).

Table 5.6: *Overview on the typical, critical failures observed for imprinted MRRs.*

Failure code	Description	Failure origin	Example
	F00-series: Imprint homogeneity		
F01	Wedge error	Particles / mold contamination	Fig. 5.21b
F02	Inhomogeneous RLT	Mold layout / polymer flow	Fig. 5.21c
	F10-series: Structure contamination		
F11	Particle on/close to structures	Handling (post-imprint)	None
F12	Particle trapped in structures	Handling (pre-imprint)*	None
	F20-series: Structure defects		
F21	Gap not resolved	Mold manufacturing / handling	Fig. 5.5b
F22	<50 % Structure missing	Air entrapment / cavity filling	Fig. 5.8b,c**
F23	>50 % Structure missing	Air entrapment / cavity filling	Fig. 5.8b,c
F24	Inhomogeneous gap	Cavity filling	Fig. 5.8d
F25	Rough sidewalls / edges	Cavity filling	Fig. 5.8e,f
F26	Non-filling	Polymer volume applied / flow	None
F27	Structure not transferred	Mold contamination	None
F28	Structure lifted	Demolding / adhesion	Fig. 5.8g-h
F29	Structure removed	Demolding / adhesion	Fig. 5.8i
	F30-series: Curing defects		
F31	Uncured structures	Curing process / parameters	Fig. 5.13b,c

* Had no influence on the residual layer. Otherwise it had to be classified as F01.

** The size determines the classification as F22 or F23.

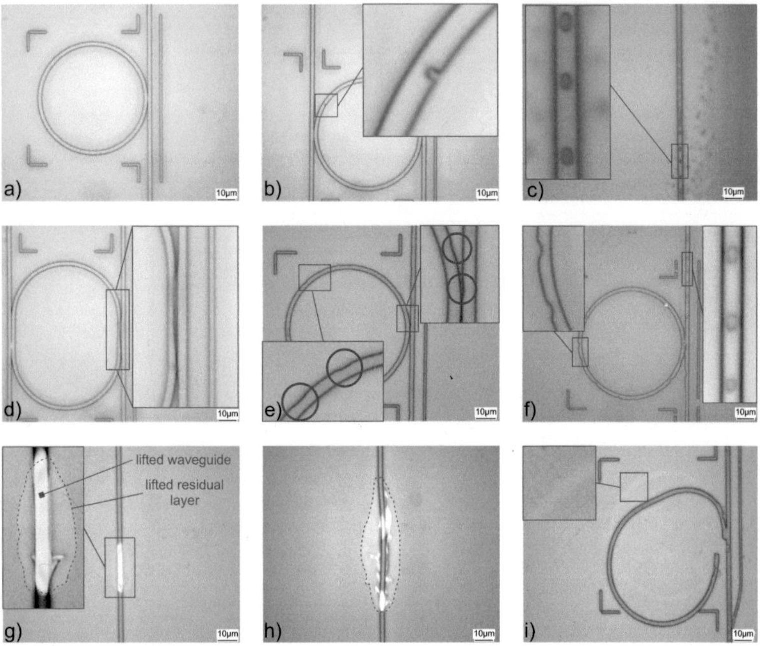

Figure 5.8: *Comparison of stamp-and-repeat imprints of a) a defect-free MRR and MRRs having typically structural failures: b) missing parts due to air entrapment; incomplete c) waveguides and d) gaps due to insufficient cavity filling; e) and d) missing parts well aligned along the waveguide sidewall probably due to insufficient cavity filling and self-assembly; g)-i) lifted structures due to insufficient substrate adhesion.*

Trapped air was found as bubbles in the waveguides (see Fig. 5.9a). Bubble like structures found outside the waveguides (see Fig. 5.9b) favored the interpretation as bubbles rather than as self-assembling defects. The bubbles outside might have been "flown-out" from the waveguide cavities. However, large defects along the waveguide sidewall or top (see Fig. 5.9c,d) indicated the presence of self-assembling or capillary effects. Different filling modes were observed in cured waveguides. For one mode, the imprint material reached the mold ceiling simultaneously close to the both sidewalls (see Fig. 5.9d). For the other mode, the ceiling was reached without contact of the imprint material to one sidewall. Again, air inside the waveguide channel and an insufficient supply of imprint material might have initiated the development especially of the latter structures.

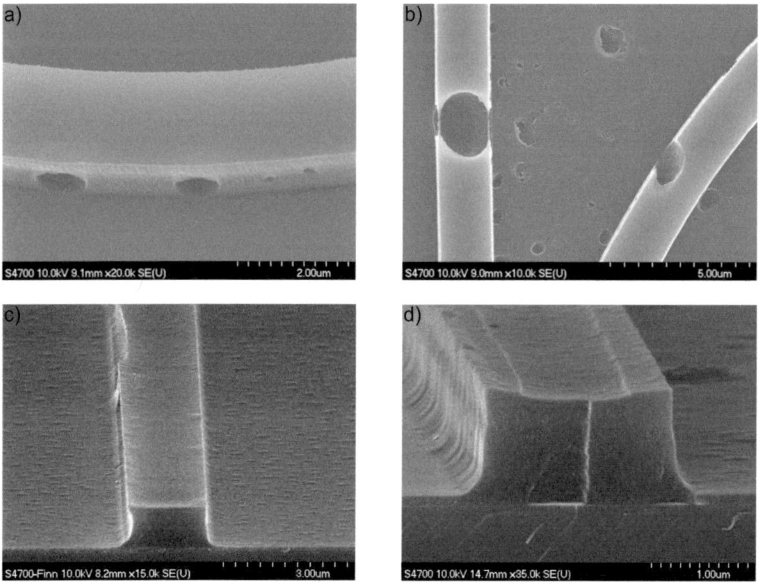

Figure 5.9: *SEM micrographs of typical filling failures of MRR waveguides due to a)-b) air entrapment and insufficient polymer flow removing the trapped air and c)-d) missing imprint material along the waveguide sidewall and the waveguide top probably due to self-assembling effects and insufficient cavity filling.*

A clear differentiation between bubble-trapping and self-assembly was difficult. Trapped air was believed to be both, a driving force for film instabilities and thus self-assembling and dewetting effects (see Fig. 5.13a) and, on the other hand, the direct result of self-assembly. In this regard, the mentioned deep (cf. Fig. 5.9c) and shallow structures (cf. Fig. 5.9f) represented different states of the removal of enclosed air by diffusion into the resist.

Yield and Relative Failure Frequency For better comparison, all structures of multiple stamp-and-repeat MRR-imprints were manually inspected and grouped according to the mentioned failures (cf. Tab. 5.6) giving an overall yield Y (see Tab. 5.7). For spin-coated

Table 5.7: *Comparison of stamp-and-repeat processes using different resist application methods.*

Parameters	Man. B		Autom. A		Autom. B		Spin-coating 750 nm B	
	Y^*	Y'^{**}	Y^*	Y'^{**}	Y^*	Y'^{**}	Y^*	Y'^{**}
Imprint fields (n)	3	3	4	4	2	2	2	2
Effective yield (AVG)	64	68	79	80	73	78	79	95
Effective yield ($\pm\sigma$)	13	14	3	3	4	3	7	5

* without mold defects due to lithography (e.g. non-resolved features)
** without mold and air trapped at the mold periphery
A: Customized, low viscous, acrylate based resist; B: UVCur21SF

films, defects were almost exclusively found at the "dead-end" of the cavities at the mold periphery (see Fig. 5.12). Thus, the yield Y was further corrected and termed Y'. The latter one does not take into account the accumulated defects at the cavities at the mold periphery because these structures can be directly designed as "dummy"-structures and do not influence the device functionality. The fact that defect free imprints can be achieved for spin-coated films and defects at the mold periphery occurred is also reflected by comparison of Y and Y' for the other resist application methods. However, such "dummy"-structures are much more difficult for binary lithography than for, e.g., direct patterning of waveguides. The major reasons for the lower yield of dispensed resists were air inclusions (F22, F23) and self-assembling (F25) (see Fig. 5.10) as discussed above.

Origin of Bubbles and Filling Defects Self-assembling was especially pronounced for the low viscous, customized imprint material compared to the more viscous UVCur21SF during automatic dispensing (cf. Fig. 5.10b). This can be explained by a higher self-assembly tendency for lower viscosities [89]. For automatically dispensed and spin-coated resist there were not as many structural defects observed as for manual dispensing. Furthermore, the defects observed for automatic dispensing were much smaller than for manual dispensing. Manual dispensing probably introduced a lot of air bubbles into the resist. For spin-coated resists almost no structural defects were observed.

The polymer flow for spin-coated films (homogeneous film) was completely different from dispensed resist patterns (droplets). Additionally, spin-coating and a subsequent soft-bake probably yielded a resist layer with only minor air inclusions. According to

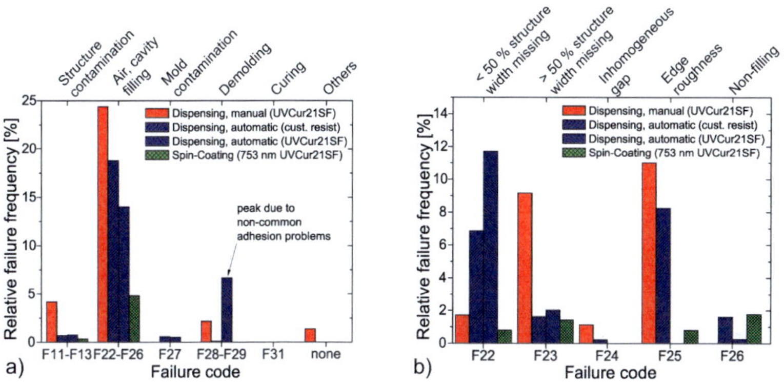

Figure 5.10: *Representative relative failure frequencies for different resist application methods: a) overview for all observed failures and b) detailed distribution for the most common failures related to air entrapment and insufficient polymer flow.*

Liang [104] the origin of bubbles for dispensed drop patterns can be either fluid pinning or droplet coalescence (see Fig. 5.11a). Due to the negative mold character, this enclosure of air between flowing fluid fronts is more probable than the pinning effect, because no pinning structures are present on the flat and unstructured mold area. A proposed removal of bubbles from mold cavities occurs when a lateral fluid flow is still present in the cavity (see "moving bubble" in Fig. 5.11b) [165]. A trapping occurs, if no lateral fluid flow is present in the closed cavity (see "trapped bubble" in Fig. 5.11b). For open cavities like waveguide channels, the bubbles can be flown out along the channel to the mold periphery to a certain extent (see Fig. 5.11c).

Both, bubble trapping in closed cavities and the transport to the mold periphery in open cavities was observed for many imprints. If a negative mold approaches ideally parallel a spin-coated film, air is always enclosed in the cavities. The fact that bubbles were not found in all closed cavities proofed that the air either diffused into the resist or that a lateral flow existed also for closed cavities. The later discussed thick residual layer in the mold center (see Sec. 5.3.4) could have act as driving force for a polymer flow inside the cavities. More bubbles and defects were found on the open cavity end (see Fig. 5.12a,b) than on the closed cavity end (see Fig. 5.12a,b). This proofed that polymer flow was present along the cavities and that bubbles were partially removed by this flow from the cavities. This lateral flow was supported by an unintended inclined imprint, which was always present and which was not possible to eliminate. Thicker resists provide more volume to flow into and then laterally in the mold cavity. Thus, thicker initial resist layers should eliminate bubbles better than thinner initial resist layers.

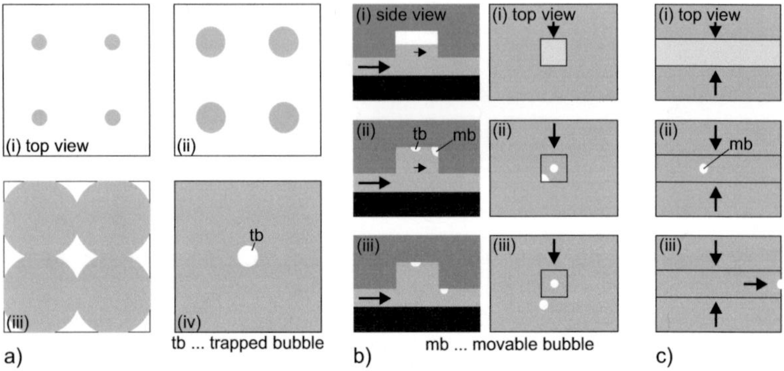

Figure 5.11: *a) Bubble trapping due to fluid front coalescence (modified from [104]). b) Bubble removal from cavities by a net fluid flow inside the cavity (modified from [165]) and bubble trapping in case of no lateral fluid flow. c) Bubble removal along an open cavity like waveguide channels.*

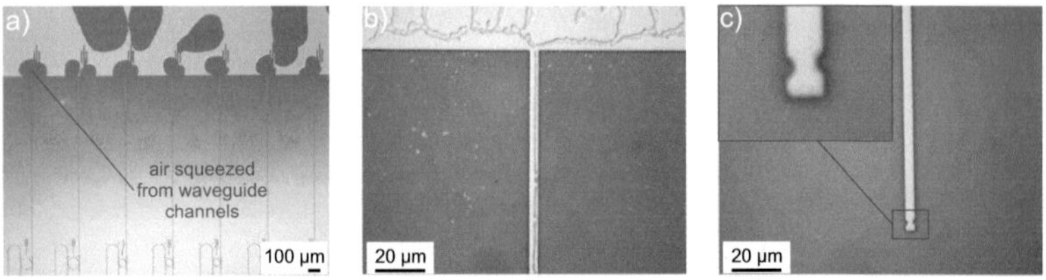

Figure 5.12: *a) The "open end" of waveguides with air squeezed along and finally out of waveguide channels at the mold periphery leading to dewetting-like effects for imprints in 350 nm spun-on films. b) Incomplete cavity filling and finger-like structure at the "open-end" for another spun-on film due to air preventing a fast filling. c) Air bubbles trapped inside a closed cavity like the "dead end" of a waveguide.*

Curing Defects due to Bubbles If the air was not completely removed from the cavities, oxygen inhibition led to uncured waveguides in cases of insufficient curing doses during imprint D_0. It was evident that such curing failures appeared only in concave structures like waveguides (see Fig. 5.13b and c) clearly demonstrating the relation to air entrapment and its influence. If the curing dose is not sufficient to overcome the oxygen inhibition, defects in the final structure will be the result. This is a particular challenge for stamp-and-repeat imprinting, where adjacent fields are permanently pre-exposed leading to a modified curing characteristic. A vacuum or partly also an inert gas environment could minimize the oxygen inhibition.

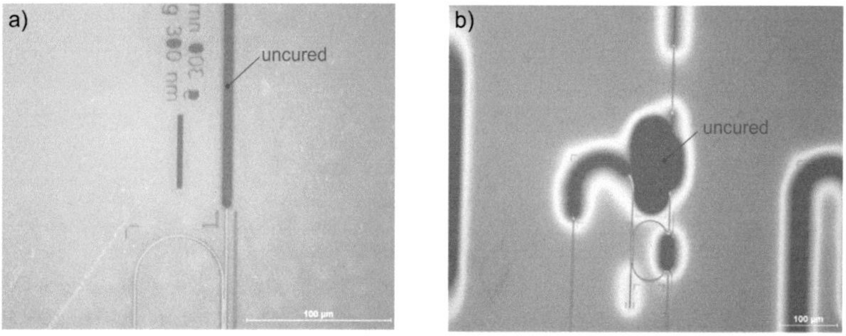

Figure 5.13: *a-b) Overview micrographs with inhibited curing due to the presence of air/oxygen within certain regions of the waveguide structures and an insufficient imprint exposure dose* D_0.

It must be noted, that this result was specific for the used mold layout. The behavior of dispensed and spin-coated films might be different for another mold layout and has to be evaluated separately.

5.3.2 Structure Filling

The defect-free and complete mold filling is crucial for directly imprinted polymer devices like MRRs. For a typical imprint profile, the final imprint force was reached by several ramps and plateaus (see Fig. 5.14). The first two ramps and the plateaus ensured a smooth contact of mold and resist. Before ramping the force to the final imprint force, a quite long plateau ensured a homogeneous contact between mold and resist. As seen later, the impact on mold filling of this plateau was only marginal and the holding time can be reduced to a few seconds.

The residual layer thickness and thus the polymer displacement from under the mold features is inverse proportional to the square root of both the pressure and the time the resist is under pressure (cf. Eq. (3.3)). For a given mold, the active imprint area is constant and the force can be used to characterize the imprint instead of the pressure.

The active area for the used mold was 10×10 mm^2 (cf. Sec. 5.2.1). The area under the force profile is the impulse applied to the resist and was used to characterize the imprint process.

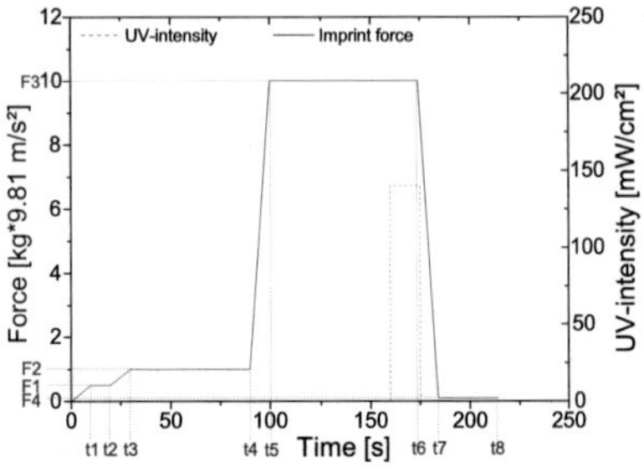

Figure 5.14: *Typical imprint profile for SR-UV-NIL.*

Capillary action plays an important role in NIL. Capillary effects can be the origin of defects and influence the structure filling [90, 166–168]. Most literature studies use vacuum environment and typical lateral structure dimensions below 500 nm or quite shallow mold cavities favoring capillary effects. In one study, for structures in the range of 5×5 µm^2 small amounts of capillary bridges were observed and found to reach a constant density for structures larger than 10×10 µm^2 [168]. However, for cavity depths larger than 80 nm no capillary bridges were observed in this study for initially 200 nm thick resist films. It was found that capillary filling of the gap between mold and substrate occurs only slowly when resist was applied at the mold periphery and is not useful for UV-NIL resist application [163]. Good estimates for the filling time can be derived from the following equation [102]:

$$t_f = \frac{\eta_0 \cdot s^2}{2p} \left(\frac{1}{h_f^2} - \frac{1}{h_0^2} \right) \tag{5.2}$$

t_f is the final time required to squeeze the resist from under the protrusion over a distance s until the resist thickness h_f is reached starting from a resist thickness h_0. Assuming volume conservation, one can estimate the time required to squeeze the cavity volume from under the mold protrusion. Useful estimates are obtained even though the influence of capillary action is neglected [102]. A T-NIL processes with viscosities of the resist during polymer flow being above 10^4 Pa·s would require filling times in the order of 10-100 min for the used MRR-layout (cf. Fig. 5.2) and the low force of 120 N. The final time t_f and thus the estimated filling time is linearly proportional to the viscosity η_0 [see Eq. (5.2)]. UV-NIL resists are assumed to be Newtonian liquids, as it is typical for low molecular weight liquids and Eq. (5.2) can be applied. Reducing the viscosity to typical UV-NIL values of 100 mPa·s, would reduce the required filling time well below 1 s. This was in contrast to the observed times required to fill the mold cavities for the used layout. Typical times for complete filling were in the order of 5 min for forces in the order of 100 N corresponding to an impulse of about 30 kNs (see Fig. 5.15). The force was of higher importance for filling than the holding time (see Fig. 5.15b, UVCur21SF layers). Longer holding time were only useful for higher imprint forces. Smaller structures were filled faster than larger structures due to their smaller cavity volume (see Fig. 5.15, 340 nm and 160 nm).

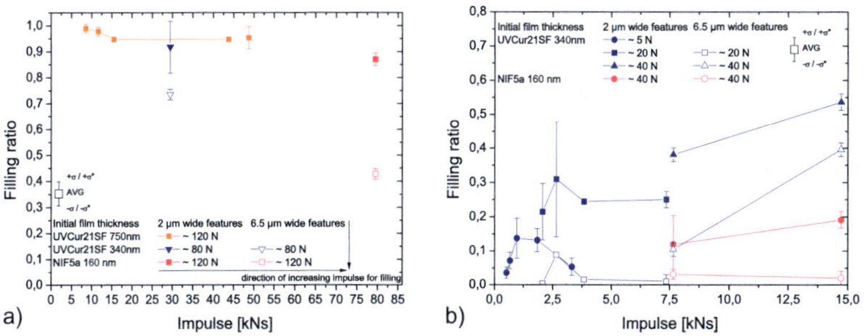

Figure 5.15: *a) Impact of the applied impulse I on the structure filling quality and b) details for low impulse values. Note: Data points connected with solid lines were obtained for constant forces and modified holding times. (σ*: estimation if data was not available)*

Local capillary filling of mold structures is often said to occur for UV-NIL under vacuum [127]. However, capillary filling seemed not to occur in the ambient-environment imprint setup of this work.

In contrast to typical imprint mold layouts, a highly negative mold was used. A quick filling of the gap between mold and resist was usually observed when the mold approached the resist surface. However, this did not include the mold cavities. For typical, initial resist layers smaller than 500 nm, the gap between the large unpatterned area and the substrate is much smaller than the cavity size. The typical contact angle was roughly 10-50° between resist and substrate and 80-90° between resist and mold and thus attractive capillary forces can be assumed for both surfaces. However, due to the larger contact angle with the mold, the higher capillary forces act between resist and substrate. As a basic rule, liquids tend to move towards smaller gaps due to capillary forces. As a consequence there was no net force pulling the resist into the mold cavities. The resist was hold back in the thin, well spreading residual layer. This was one reason why complete mold filling was much more difficult for thin NIF5a layers compared to thicker UVCur21SF layers (cf. Fig. 5.15).

As a consequence, the initial layer thickness should be in the order of the height of the mold cavities. This results in an initial film thicknesses which is much larger than used for the typical imprinting of nano-sized features. In case of a comparable size of cavity width and residual layer, a quicker filling is assumed due to smaller capillary forces pulling the resist back into the residual layer gap, even though the mold-resist and substrate-resist contact angles are still quite different. Dummy-fill structures will reduce the amount of unpatterned areas holding the resist back. They will also take up additional resist from the thick initial layer. The amount of dummy-structures can be adjusted based on resist volume-conservation during imprinting. Only a thin residual layer should be left after contacting the resist with the mold.

5.3.3 Neighboring Exposure Effects

The used standard mold (see Fig. 5.1) was mounted face down in the imprint system. The UV-LED array (λ=365$\pm$25 nm, max. D_0=250 mW/cm^2) was located above the mold and the UV-light passed through the mold. The active imprint area (mesa) was much smaller than the overall mold. This led to the typical imprint situation where adjacent imprint fields were already exposed to UV-light before being imprinted (see Fig. 5.16a). The accumulated dose in these adjacent fields depended on the imprint field density and D_0. For 15$\times$15 mm^2 imprint sites stitched together without gap, the accumulated dose was calculated exemplary and expressed as multiples of the imprint dose D_0 (see Fig. 5.16b)[6].

The model calculated the relative dose for each imprint field based on the area overlap of the mold with the respective imprint fields on the wafer. Relative doses were than summed-up for each imprint field. It was assumed that the relative dose is equally distributed in the imprint fields even for fields having only a partial overlap with the mold (cf. marked field in Fig. 5.16). A sharp border due to partially covered fields was never observed because the dose was accumulated for multiple exposures. Furthermore, the resist flows across the imprint field during imprinting and the angular intensity distribution was rather broad. All together led to a blurring of such a sharp border and the accumulated dose can be well estimated with the mentioned model. The accumulated dose in the first imprint field was Zero and the imprint sites per row were symmetrical distributed on the wafer but not in an equal amount per row. Thus, the maximum of the accumulated dose was not perfectly located in the wafer center and not symmetrically distributed over the wafer.

The force required to detach the mold from the imprint, also termed demolding force F_d, was recorded for all imprints. The profile and the absolute value of F_d was found to be a reliable indicator for the substrate-resist-mold interaction. Other groups for example used F_d to determine the influence of fluorine incorporation into resists on the adhesion [169]. The higher F_d the higher is the adhesion of the resist to the substrate and to the mold. Low F_d-values or deformed force profiles often indicated an insufficient substrate adhesion (cf. Fig. 5.17a for a typical F_d profile). Significant effects of ASL-degradation on F_d were not found throughout this work.

[6]The active imprint area itself was only 12$\times$12 mm^2 giving an inter-imprint spacing of 3 mm.

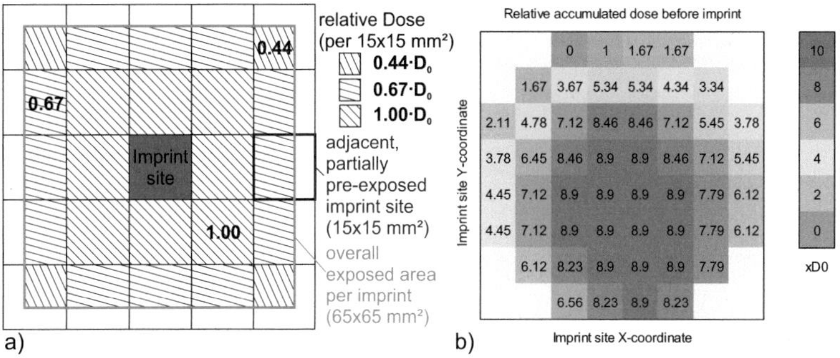

Figure 5.16: *a) Relative doses and exposure situation of the 65×65 mm² mold illumination area overlapping several 15×15 mm² imprint fields and b) accumulated dose before imprinting the respective imprint site expressed as a multiple of the imprint dose D_0.*

Degradation of TCSP-MA used as Adhesion Promoter If drop-on-demand dispensing (DOD) was used, a decreasing adhesion was observed for adjacent imprint fields when trichlorosilylpropyl-methacrylate (TCSP-MA) was used as adhesion promoter (see Fig. 5.17). Three possible reasons were isolated: dehydration bake, pre-exposure and antisticking layer. To ensure a good adhesion, the complete wafer was dehydration-baked at 160 °C in the imprint system right before imprinting. The accumulated dose on the adjacent imprint fields was 0 for the first and constantly 4.02 J/cm² for all subsequent imprints due to a low imprint field density on the wafer.

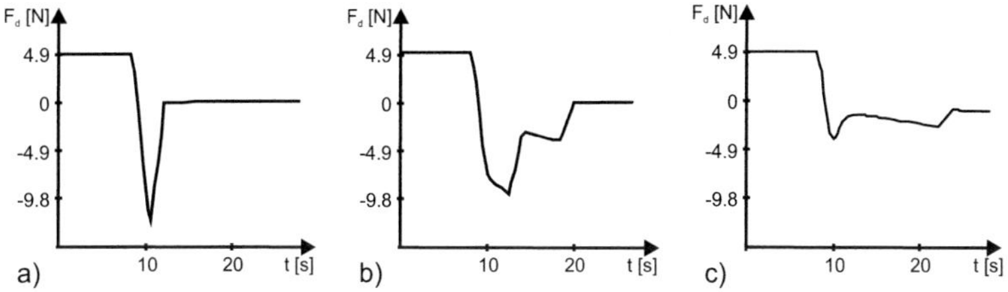

Figure 5.17: *Demolding force F_d for a) 1st and 2nd imprint (= good adhesion), b) 3rd imprint (= decreased adhesion) and c) 4th imprint (= adhesion failure) for imprinting with UVCur21SF on degrading TCSP-MA.*

The ASL was ruled out as reason because the first imprint was successful and only a slow ASL-degradation during imprinting was found (cf. Sec. 4.1.3). The accumulated dose was constant and only for the first imprint different. Thus, degradation due to UV-exposure might have been present but not the only reason. Also dehydration bake seemed to have a certain influence. This was surprising because TCSP-MA was baked for 5 min at 160 °C directly after spin-coating without any visible influence on adhesion even with a second dehydration-bake before the first imprint. Even more surprising, no dehydration bake was done before the last imprint, showing the worst adhesion. A combined influence of UV-exposure and dehydration bake seemed to be most probable. For TCSP-MA, the siloxane binding to the substrate and the alkyl chain should be thermally stable well above 160 °C. The dehydration bake seemed to have degraded the pendant methacrylate groups of TCSP-MA (see Fig. 5.18b) as also observed by another group [170].

To further elucidate the degradation, the influence of UV-exposure and dehydration bake was investigated systematically in cooperation with *L. Bo* [132]. The water contact angle was used as indicator for surface modifications. The contact angle of the spin-coated TCSP-MA films were significantly different from surfaces dip-coated in the same 1 mM TCSP-MA solution in isooctane for more than 20 min. The contact and reaction time during spin-coating was about 30 s and hence much shorter than for dip-coating. Thus, the TCSP-MA layer was expected to be less dense. However, a modification of the surface by spin-coating was present. Even though the surface modification was not very obvious from the contact angle values, compared to native SiO_2, the spin-coated TCSP-MA films represented efficient adhesion promoter layers. Multiple, subsequent UV-exposures (each 15 s at 200 mW/cm^2 and 365 nm) significantly decreased the water contact angle (see Fig. 5.18a). For a single exposure a drop of 5°was observed. After 75 s exposure, the contact angle dropped for about 10°. Such a clear decrease was not observed for multiple, subsequent dehydration bakings (each 5 min at 160 °C). However, storage of the wafer under standard clean room environment and a subsequent dehydration bake clearly decreased the contact angle with water. This decrease was even more pronounced for extended storage times (compare 1st bake after immediate bake and after 2d and 9d storage). After storing the wafers without further treatment, always an increase of the contact angle was observed. This recovery was attributed to the normalization of the concentration of hydroxyls or other surface active groups.

In summary, dehydration baking of freshly TCSP-MA coated wafers immediately after coating was not as critical as the pre-imprint UV-exposure of adjacent imprint sites. Thus, UV-exposure seemed to be the major contribution to the adhesion failures observed for the above mentioned subsequent imprints. Knowing the behavior of the adhesion promoter during UV-exposure and dehydration baking is fundamental for successful DOD used with SR-UV-NIL.

One solution to minimize the pre-exposure of adjacent fields would be coating of the non-active regions with a light blocking layer (cf. "inactive surrounding" in Fig. 5.1a). However, this makes processing more complex. The chromium coating applied as light blocking layer was found to be not long-term stable against mold cleaning processes (acidic piranha, oxygen plasma). Thus, the use of chromium requires either a recoating, which is a source for new defects and particles on the mold or the development of an

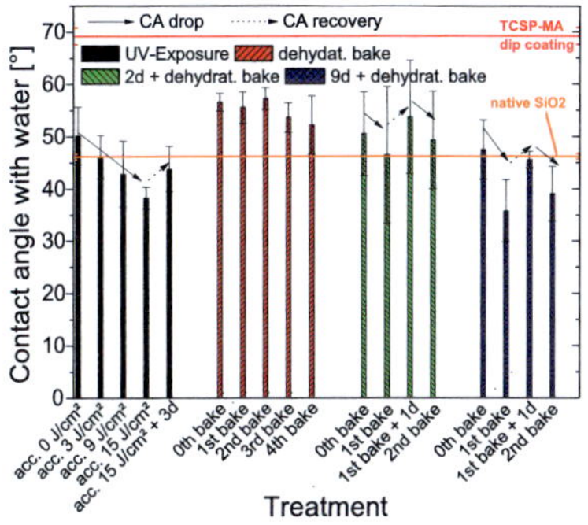

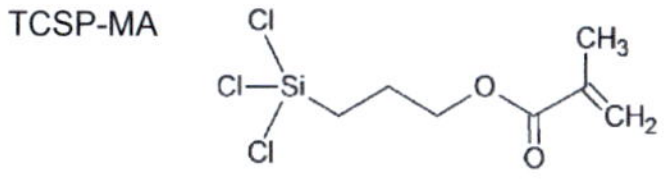

b)

Figure 5.18: *a) Water contact angles for 6 in wafers spin-coated with 1 mM TCSP-MA in isooctane (soft-baked for 5 min at 160 °C) and different post-treatments. b) Structure of the TCSP-MA molecule.*

improved chromium coating or an alternative material, which were both beyond the scope of this work. Also an aperture could be inserted into the exposure system.

Resist Curing The degree to which the resist is cured depends on the availability of photoinitiator radicals [147]. Oxygen inhibition is for example well known for acrylate based systems (cf. Sec. 3.4.4). Normally, the initiator concentration is more than sufficient to cure the resist even in the presence of oxygen, provided that the dose D_0 is high enough. Multiple exposures lead to a reduced amount of available photoinitiator molecules (see Appendix B)[147]. A direct measure for the degree of resist cure during imprinting would require an in-situ measurement of the monomer conversion ratio. Because this is difficult to implement in standard imprint systems, the demolding force F_d was used as an easy to access, indirect measure for the degree of resist curing. To determine the influence of the pre-exposure on the resist curing, F_d was plotted against the accumulated dose. For a quantitative analysis, F_d over the accumulated dose D was modeled as a symmetric, sigmoidal dose-response curve:

$$F_d = A_1 + \frac{A_2 - A_1}{1 + 10^{(D_{50} - D)p}} \tag{5.3}$$

This four-parametric model is defined by the upper limit A_2, the lower limit A_1, the D_{50}-value and the slope p (see Fig. 5.19a). A_1 described the maximum demolding force observed for the unexposed resist. A_2 was the minimum demolding force observed for large values of D. D_{50} was the accumulated dose required to reduce F_d to 50% of the sum of the upper and lower limit value. p is defined as the slope of the dose-response curve. It is a measure for the contrast of the curve.

High accumulated doses: For very high doses, a stable, minimal F_d of about 9.81 N was observed. Because not all measurements gave enough data for high accumulated dose values and the A_1-value was expected to be mostly independent from all other imprint parameters, A_1 was set constantly to 9.81 N for all fitted dose-response curves. A_2 was set to the F_d value observed for the not pre-exposed resist. This left p and D_{50} as two degrees of freedom to be fitted in the four-parametric model. Four different resist thicknesses were compared for a wide range of accumulated doses (see Fig. 5.19b). The data was fitted for the available measurements (solid line) and then extrapolated for higher doses for an improved visualization (dashed line).

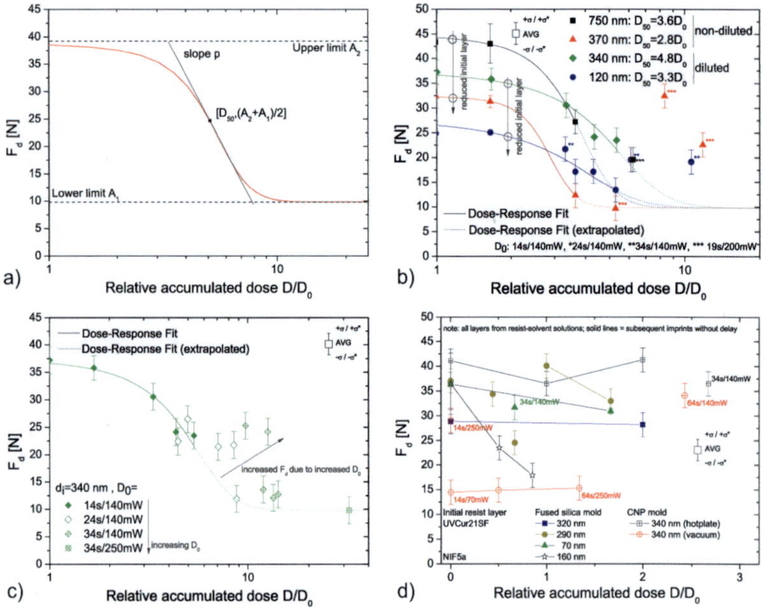

Figure 5.19: *Relation of the demolding force F_d on the accumulated dose D for stamp-and-repeat UV-imprinting on a 6 in wafer in a 15×15 mm² wafer grid: a) model of the used doses-response-fit, b) behavior for large D (for complete curve of 340 nm see c), c) details for large D showing the influence of increased imprint doses D_0 and d) behavior for small D. (σ*: estimation if data was not available)*

The demolding force of the first imprints was more a less constant or decreased only slowly until D reached about $1.5 \times D_0$. This was equivalent to about three to five subsequent imprints on a densely populated wafer (cf. Figs. 5.16b, 5.21a). The constant F_d in the beginning indicated the presence of sufficient initiator molecules, even after several pre-exposures. For further increasing D-values, F_d dropped significantly. In addition, a general decrease of A_2 was found for decreasing initial film thickness of the spin coated resist (see Fig. 5.19b). The reason for both the significant drop of F_d and the dependence on the film thickness was the imprint field sidewall. An imprint field sidewall or edge bead (cf. Fig. 5.20) is formed by the imprint resist squeezed out of the active imprint area to the edge of the mesa. Depending on the initial film thickness and the imprint parameters, the typical edge bead hight was about 10-50 µm. The thinner the initial film thickness, the smaller was the edge bead and thus its contribution to the demolding force. The large drop in F_d after several imprints indicated a large overall influence of the edge bead. This F_d-drop mainly originated from a non-cured edge bead. Thus, A_1 can be seen as a measure for the adhesion originating only from the cured imprint structures.

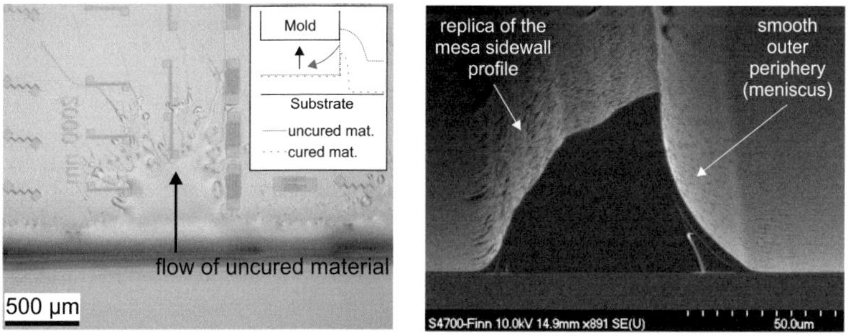

Figure 5.20: *a) Collapse of the edge bead at the mold periphery due to oxygen inhibition (mrUVCur21SF, 60 s at 250 mW/cm² at 365 nm) and b) typical edge bead profile.*

Besides a non-cured edge bead, also curing defects in structures were observed for high accumulated doses (see Fig. 5.13b and c). First defects started to occur from about the 6th and 7th imprint of a densely populated wafer (see Fig. 5.21a). This corresponded to about 3-6 × D_0 (or 5.9-11.8 mJ/cm^2) and, thus, to the calculated D_{50} values (see Tab. 5.8). Hence, D_{50} can be used as the limit for the accumulated pre-exposure dose at which curing defects start to appear within structures. An increase of the current imprint dose D_0 will allow a defect free curing even for large accumulated doses. It was proofed that increasing D_0 raised F_d compared to lower D_0-values (see Fig. 5.19c) for the same accumulated dose D. However, increased D_0 values will even more increase the accumulated pre-exposure dose in the next imprint field and, hence, reduce the amount of possible subsequent imprints on pre-exposed, adjacent fields.

Two different resist behaviors were observed, being characterized by different p-values (see Fig. 5.19b, Tab. 5.8). The slope seemed to be independent of the initial resist thickness. The only difference between group A formed by the 750 nm and 370 nm film and the group B formed by the 340 nm and 120 nm film, were the film preparation. Group A was spin-coated from undiluted UVCur21SF whereas group B was spin-coated from UVCur21SF diluted by the recommended solvent T1070 (micro resist technology GmbH). Apparently, the solvent had an influence on the curing, even though the solvent should have been removed to a large extend during 2 min softbake at 90 °C. Two effects were observed for the diluted films. The initial F_d is increased and the slope is reduced compared to undiluted films.

Table 5.8: *Parameters of the dose-response fit from Fig. 5.19b for different initial thicknesses of UVCur21SF. Note: A_1 was always set to 9.81 N and A_2 was set to the respective F_d-value for not pre-exposed resist.*

Resist thickness [nm]	A_1 [N]	A_2 [N]	D_{50} ($\pm\sigma$) [$\times D_0$]	p ($\pm\sigma$) [cm²/J]	corr. R^2
750	9.81	45.3	3.6±0.03	-0.58±0.04	0.998
370	9.81	32.4	2.8±0.03	-1.16±0.03	1.000
340	9.81	38.9	4.6±0.23	0.28±0.05	0.944
120	9.81	29.4	3.3±0.17	-0.33±0.03	0.964

Low accumulated doses: As mentioned above, low accumulated doses were characterized by almost constant F_d values (see Fig 5.19d). However, some interesting effects were observed. Comparable demolding forces were present for the 320 nm, 290 nm and one 340 nm film due to their comparable initial thickness and preparation. Whose average F_d was 35.3 N within a 1σ range of 5.9 N. This was well comparable to the data observed for similar thicknesses before (cf. Fig. 5.19b). Differences in F_d were ascribe to different resist-solvent ratios and statistical variations. The 70 nm film had also comparable F_d-values to the before mentioned films. Actually, a decrease was expected due to the thinner initial layer and a smaller edge bead. The fact that this was not the case was attributed to the largely different resist-solvent ratio for this film compared to the other films. As conclusion, spin-coating of diluted films changed the mechanical properties and the curing behavior of the film and in particular increased F_d.

Increasing the D_0 had no influence on F_d for low accumulated doses. For example, for the 70 nm film 36.4 and 35.3 N were measured for 14 s and 34 s curing with the same intensity (data not presented), respectively. For the one 340 nm film 39.2 and 43.2 N were measured for 14 s and 34 s curing (data not presented), respectively. This was well within the statistical variation and revealed that below D_{50} enough initiator molecules were present to achieve a high degree of curing. Notably, the two 340 nm films showed a very different behavior. The only experimental difference was, that the high F_d film was soft-baked for 2 min at 90 °C on a hotplate and the low F_d film was baked for about 30 min at 80 °C in a vacuum-oven at 10-30 mbar. The latter process was assumed to removed the spin-coating solvent to a greater extend from the resist film. Additional influences on the resist like stoichiometrical changes due to evaporation of resist components are also possible. As a result, the mechanical and curing behavior of the film was significantly different. It seemed, that the edge bead might not be fully cured for this film, because an increase of the curing dose lead to a significant increase of F_d right to the normally observed range of F_d for comparable films in two cases [cf. Fig. 5.19d, 340 nm (vacuum)]. This corresponded to a reduced UV-sensitivity. Additionally, if subsequent fields were imprinted without time delay, the resist seemed to require a higher curing dose than a site which was pre-exposed but which was not immediately imprinted and where the resist had more time to recover. The same was already observed for larger accumulated doses (cf. Fig. 5.19b, 370 nm thick film, $D>5\times D_0$). In addition, no effect was found for cured and uncured residual layers on F_d when hybrid molds were used (cf. Sec. 5.4) because the edge bead dominated F_d. In comparison with another resist, NIF5a (Asahi Glass Company[7]), UVCur21SF was less sensitive to pre-exposure. For NIF5a, already small accumulated doses lead to a steep F_d-decrease. This could indicate for example a lower photoinitiator concentration.

In summary, F_D was found to be a useful measure for the degree of imprint material curing. Spin-coating with solvents influenced the curing behavior and the final material properties. The success of SR-UV-NIL depends on the type and concentration of the photoinitiator in the imprint material. The tested UVCur21SF can be used for a medium dense wafer population.

5.3.4 Residual Layer Homogeneity

The imprint curing dose was adjusted for each imprint to ensure correctly cured structures and a densely patterned 6 in wafer. Uncured UVCur21SF was easily removed from the wafer by an intensive rinse or spin-drying with isopropyl alcohol. A reproducible, inhomogeneous residual layer thickness (RLT) was observed for typical SR-UV-NIL imprints on the 6 in wafer (see Fig. 5.21). The difference between the residual layer center and the periphery was quite constant with about 200-250 nm showing a dome-like RLT distribution (see Fig. 5.21c). Only one imprint with 50 nm RLT was achieved but could not be reproduced. Particle defects were observed as wedge errors (see Fig. 5.21b) and clearly differed from the observed inhomogeneity. It should be remembered that the required impulses for sufficient mold filling were about 10 kNs, 30 kNs and 80 kNs for about 750 nm, 350 nm and 150 nm thick initial layers, respectively (cf Sec. 5.3.2).

As a "rule-of-thumb", a residual layer thinner than 5 % of the waveguide height was envisaged. Thus, the aim was to reach an RLT below 50 nm. This was comparable large for binary lithography specifications but still a well acceptable value for direct patterning of micrometer high features. RLT specifications must be rethought for direct imprinting approaches. The main results found for systematic investigations on the influence of initial film thickness, imprint holding time and imprint force are discussed in the following. The initial mold contact was defined as wetting of the mold, which was easily observed due to different optical properties of the mold-air and mold-resist interface. The initial mold contact occurred always in one corner or along the edges of this particular corner. An initial contact in the mold center was never observed. The developed centralized inhomogeneity thus indicated a significant influence of the imprinting process.

[7]Thanks to Dr. Holger Schmitt at Fraunhofer IISB for the provided resist sample.

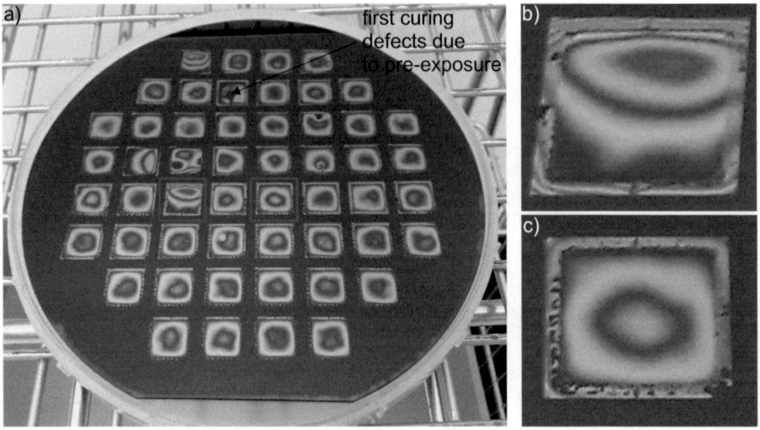

Figure 5.21: *a) 6 in wafer imprinted with a structured mold by stamp-and-repeat UV-NIL in a 370 nm thick UVCur21SF and with a typical b) wedge error originating from particles between mold and wafer and c) an inhomogeneous residual layer distribution across the imprint due to improper polymer flow.*

Observed RLT-profile Characteristics The peak of the dome like RLT-profile (see Fig. 5.22) was always well centered within the imprint field. The peak was further mostly symmetric to the imprint center and well reproducible. Shifts of the peak away from the imprint center were observed for certain imprints (see Fig. 5.22b: 44.5 kNs, Fig. 5.22c). When the peak was shifted away from the center, the RLT-profile became asymmetric. This is obvious for a peak which is not perfectly centralized but which has similar RLT-values at the periphery. Such a peak will be difficult to achieve with a bow of the mold or the substrate. Such a bow was assumed to be always symmetrical due to the imprint setup.

Influence of Initial Thickness No significant reduction of the central RLT compared to the initial layer was achieved for initial layers thinner than 300 nm. Polymer flow to the periphery of the active imprint area seemed to be minimal for such thin initial layers (cf. Fig. 5.22a-c). This fits the before observed phenomena, that the resist is hard to be squeezed out of very thin resist layers. One reason might have been the capillary retraction into the gap. According to the conservation of volume, the initial thickness should be 65 nm to reach the specified 50 nm RLT for the current layout. Filling from such thin layers is almost impossible within reasonable timescales as discussed before (cf. Sec. 5.3.2). Dummy fill structures are required to reduce the unpatterned area as well as capillary forces and allow for larger initial layers.

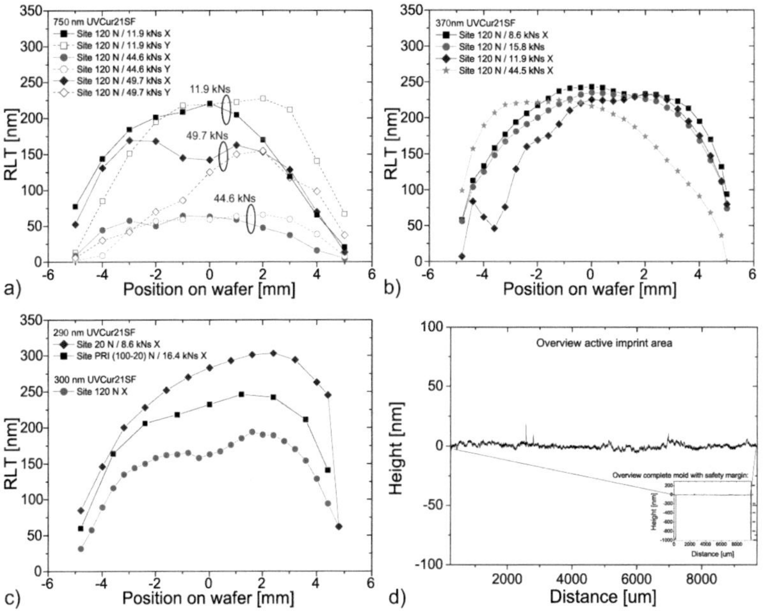

Figure 5.22: *Residual layer profile for different UVCUr21SF spin-coated films of a) 750 nm, b) 370 nm and c) about 300 nm. The unloaded mold had no visible bow as observed with d) stylus profilometry.*

Influence of Holding Time and Imprint Force Long holding times at constant imprint forces reduced the central RLT (see Fig. 5.23a,b). The specified 50 nm were not reached in reasonable times below 10 min per imprint. Higher imprint forces increased the central RLT and reduced the peripheral RLT and thus the spread between center and periphery (see Fig. 5.23d). Longer holding times can reduce the spread and compensate the influence of the higher forces only partially. All these results indicated a deformation of the imprint partners mold or substrate. However, already at extremely small impulses lower than 1 kNs achieved with forces of less than 5 N (equals 50 kPa), a spread of 100 nm was already present. Much higher forces increased the spread only by 50 % to about 150 nm. Additionally, the wafer was chucked by an about 10 mm thick silicon-carbide vacuum plate. The mold itself was 6.35 mm thick and vacuum chucked by a 25 mm thick fused silica plate. The complete imprint system worked in a parallel-plate pressing regime. Again a bow of the mold, the imprint system or the substrate cannot account for this RLT-inhomogeneity. Planarity failures in the imprint system were also excluded.

An inhomogeneous hydrostatic pressure distributions was reported during the onset of polymer flow under imprint features [171]. The pressure was found to be up to three times higher under the central part of quadratic imprint features and dropped towards the feature periphery. This study reported further that the polymer layer thickness (flow channel) primarily determines the pressure increase. Due to the low pattern density, the complete imprint mold used in this work acts basically as one single, quadratic imprint feature. The pressure concentration in the mold center led to strain in the imprint mold and the substrate. A stress of 1.2 MPa can be assumed from a 120 N imprint with a 100 mm^2 active imprint area. A rough Hookean estimate gives a strain of 6.5 ppm for the silicon substrate (Young's modulus: 185 GPa) and 16 ppm for fused silica (Young's modulus: 75 GPa) for the 1.2 MPa. The silicon wafer will be only deformed about 4 nm. However, the 6.35 mm thick fused silica mold will be deformed about 100 nm under these conditions. Assuming a pressure increase up to three times in the mold center results in a deformation of about 300 nm. Those values and the hydrostatic pressure profile under a quadratic pad derived in [171] fit the observed residual layer inhomogeneity. In conclusion, high forces led to deformations in the imprint system. A solution for the inhomogeneous RLT is the adjustment of the mold layout and the initial film thickness. However, no explanation for the RLT-inhomogeneity at low imprint forces was found.

Residual Layer Removal Effects on Structure Homogeneity The residual layer can be removed with O_2-RIE. This transfers the RLT inhomogeneity into a structure height inhomogeneity. The average structure height over the $10 \times 10\,mm^2$ imprint field measured by profilometry (see Fig. 5.24a) was comparable to the inspection of the mold (see Tab. 5.3) and waveguide cross sections (see Tab. 5.4). Visually, an inhomogeneous RLT could be seen along the line ranging from field A23 to field D10 (cf. Fig. 5.24c). For this certain imprint, the RLT was about 180 nm in this area and only about 10 nm around the fields A1 and D25. Removing 90 nm of polymer resulted in already reduced waveguide heights in the low RLT corner fields compared to intact waveguides in the thick RLT region (see Fig. 5.24b). Etching again about 90 nm removed the residual layer completely

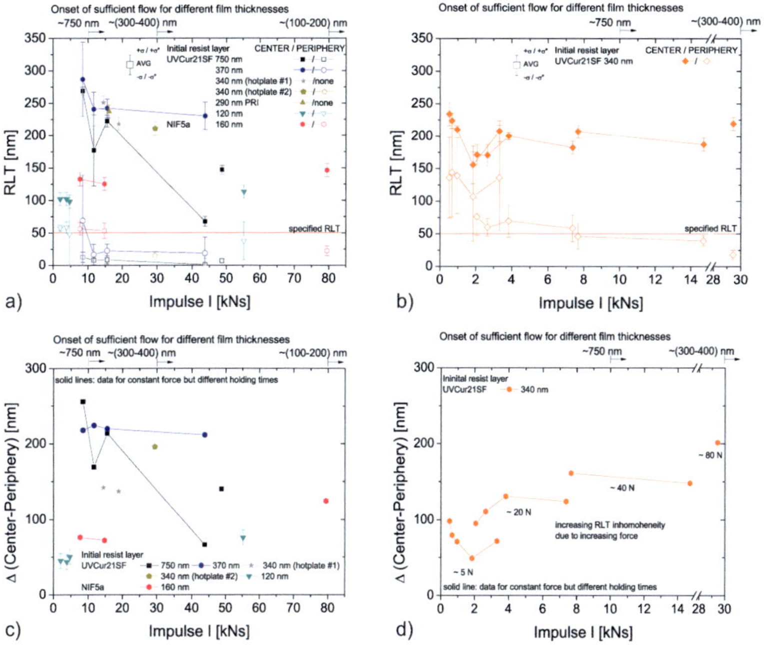

Figure 5.23: *Residual layer thickness measured in four imprint corners (periphery) and the imprint center (center): a-b) comparison of the periphery and center measurement as well as c-d) difference between center and periphery as measure for the RLT-inhomogeneity for several spin-coated layer thicknesses. Note: Solid lines are drawn for data with constant forces and increasing holding times.*

except for a very small region around field B22 (see Fig. 5.24c).

a)

Height (nm)	25	24	23	22	21	20	19	18	17	16	15	14	13	12	11	10	9	8	7	6	5	4	3	2	1
A	968	975	964	964	973	960	969	989	982	969	967	970	969	969	968	955	968	968	973	966	968	963	969	968	929
B	971	970	970	1015	978	969	969	967	962	979	959	970	978	963	981	969	969	966	969	962	964	972	999	963	978
C	975	967	972	978	970	975	985	982	986	957	973	973	973	973	972	971	972	967	966	971	972	964	966	966	959
D	974	969	970	1000	998							976	976	973	981	968	974	972	964	964	970	966	1006	990	976

AVG [nm]	± σ [nm]	Min [nm]	Spread [nm]	Max [nm]
972	11	929	86	1015

Scale: 750 773 795 818 840 863 885 908 930 953 975 998 1020 [nm]

Location of AVG

b)

Height (nm)	25	24	23	22	21	20	19	18	17	16	15	14	13	12	11	10	9	8	7	6	5	4	3	2	1
A	963	963	970	979	980	970	972	976	980	965	960	946	960	951	943	932	923	911	901	905	902	921	869	877	871
B	969	974	977	970	979	971	973	957	975	967	971	952	993	933	957	952	957	958	954	944	931	916	898	891	880
C	953	934	979	967	990	977	968	959	975	967	979	976	974	970	959	973	973	979	971	964	954	928	914	888	871
D	946	924	948	972	995							984	972	979	993	977	973	976	963	983	949	923	911	910	872

AVG [nm]	± σ [nm]	Min [nm]	Spread [nm]	Max [nm]
951	32	869	126	995

Scale: 750 773 795 818 840 863 885 908 930 953 975 998 1020 [nm]

Location of AVG

c)

Height (nm)	25	24	23	22	21	20	19	18	17	16	15	14	13	12	11	10	9	8	7	6	5	4	3	2	1
A	844	853	882	881	877	878	868	857	873	848	851	839	822	836	825	813	803	792	756	784	787	787	780	771	758
B	878	906	987	982	966	939	913	904	896	869	865	844	829	828	835	834	842	842	838	820	816	803	796	773	764
C	816	832	874	883	898	903	905	891	927	902	898	891	875	888	878	872	867	869	861	853	839	824	793	780	756
D	801	809	840	859	850							895	905	905	896	898	885	878	862	849	843	815	795	780	771

AVG [nm]	± σ [nm]	Min [nm]	Spread [nm]	Max [nm]
850	50	756	231	987

Scale: 750 773 795 818 840 863 885 908 930 953 975 998 1020 [nm]

Location of AVG

Figure 5.24: *Development of the waveguide height distribution of an MRR imprint with a $10 \times 10\,mm^2$ area and an inhomogeneous residual layer distribution measured by profilometry: a) directly after demolding (max. RLT 180 nm), b) after 1st O_2-RIE (90 nm polymer etch, max. RLT left: 90 nm) and c) after 2nd O_2-RIE (90 nm polymer etch, max. RLT left: 0 nm). O_2-RIE parameters: 60 sccm O_2, 75 mTorr, 68 W HF, 342 V self bias*

The transfer of the initial RLT-inhomogeneity can be clearly seen in a growing spread between the height of the highest and lowest waveguides. The final spread is well comparable to the typical difference observed for the RLT imprinted with the current mold layout. The initial spread was attributed to measurement errors and possible air inclusions on the waveguides top, which could not be differentiated from the total waveguide height with standard 2.5 µm profilometry cantilevers.

5.4 Hybrid UV-NIL Technique (CNP)

Avoiding the residual layer completely would by the best solution to reduce the RLT-inhomogeneity, the post-processing steps and their structural impact after imprinting. This can be realized with a combined nanoimprint and photolithography (CNP) technique and with hybrid nanoimprint molds (HNMs) having light blocking metal layers (see Figs. 5.3c-d,5.25a). The uncured material can be removed by solvent rinse or evaporation of low molecular weight components. As light blocking layer, the chromium hard mask from mold etching was used (cf. Sec. 5.2.1).

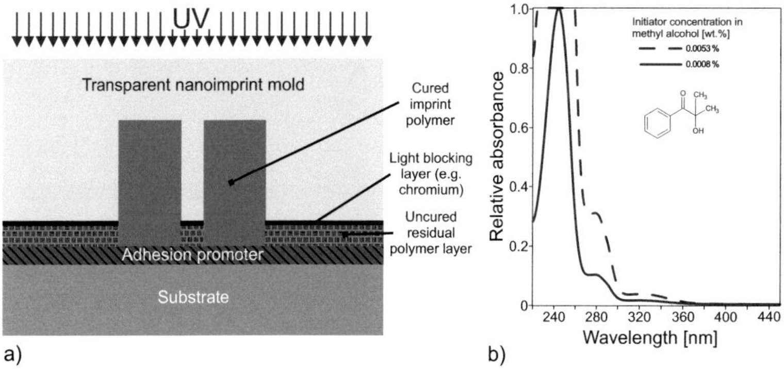

Figure 5.25: *a) Principle of a HNM for residual layer free replication of microring resonators and b) spectrum of the used UV-photoinitiator 2-hydroxy-methyl-propiophenone [172].*

5.4.1 Imprinting Results

A small strip of the chromium layer at the mold periphery was removed and a sharp cut-off of the residual layer was observed along the border between the intact and this removed chromium coating (see Fig. 5.26a). This demonstrated the light blocking ability of the chromium layer and the residual layer free patterning. Free standing waveguides without a residual layer were achieved on a native oxidized silicon substrate (see Fig. 5.26b). Microring resonators were also replicated residual layer free in customized imprint material (see Fig. 5.26c). Only a very small and localized defect was present in the gap region being about 350 nm wide (see Fig. 5.26d). This cured region probably resulted from a damaged chromium layer in the gap region or from light propagation under the chromium within the 3 µm thick SiO_2 layer. For later discussion, it is important to note that the exposure on the native and the 3 µm thick oxide was done for 30 s at $200\,mW/cm^2$ ($6\,J/cm^2$).

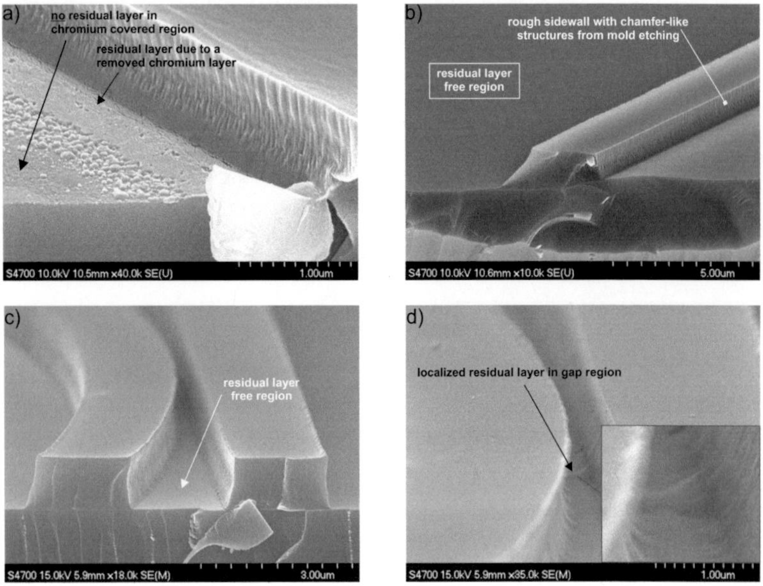

Figure 5.26: *Imprints with a HNM in the customized resist formulation A+A: first, on a silicon wafer with native oxide showing a) a roughly 70 nm residual layer at the imprint periphery, where the chromium was removed from the mold as well as b) a free-standing waveguide; secondly, on a silicon wafer with 3 µm tick SiO_2 layer: c) microring resonator gap having no visible residual layer in the 350 nm gap region except for d) a local defect arisen from a mold damage or light propagation under the chromium covering.*

5.4.2 Chromium Layer Degradation

One challenge for CNP-molds was the degradation of the chromium layer due to multiple mold cleanings and constant use for about eight months. The initial chromium thickness was about 50-55 nm (see Fig. 5.27a). After a long-term use, the chromium thickness was reduced to about 20-25 nm and clear chromium removal starting at the mold periphery could be observed (see Fig. 5.27b).

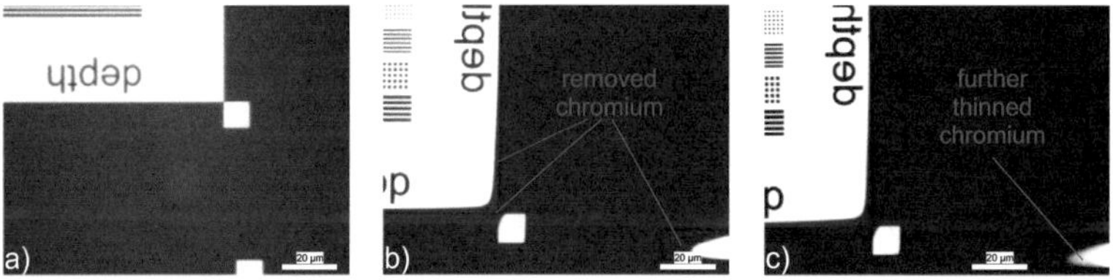

Figure 5.27: *Removal of Cr (bright areas) from the mold edges visualized by comparison of a) an imprint mold as received from the vendor, b) molds after about eight month of frequent use (equals 10-15 cleaning and ASL-recoating steps) and c) after two additional, intensive, hot acidic piranha cleaning steps (in total about 1 h at 120 °C acidic, 2:1 piranha solution).*

A removal and damage of the chromium on HNMs was observed for single acidic piranha cleanings (see Fig. 5.27c). However, a thinning (vertical removal) of the chromium could not be measured with profilometry due to the limited vertical resolution. The lateral removal rate for a very intensive piranha cleaning for 95 min at 120 °C was about 0.4 µm per cleaning. Such a high lateral removal rate was not observed for the RIE-process (50 sccm O_2, 10 Pa, 66 W HF, approx. 380 V self bias, 15 min) (see Tab. 5.9).

Table 5.9: *Lateral removal rate of the Cr-layer on HNM after hot acidic piranha cleaning and Ar/O_2-RIE. (Note: Large σ-values are caused by the measurements with an optical microscope.)*

Procedure	Cr-removal rate	
	AVG [µm]	$\pm\sigma$ [µm]
Piranha	0.4	0.2
Plasma	0.0	0.1

Chromium is also a typical hard-mask material for dry etching. The impact of RIE-processes is assumed to be lower than the chemical attack of hot acidic piranha. Short, moderate temperature acidic piranha and short RIE-processes are used during mold activation for antisticking layer coating. Those processes have only a small, acceptable impact on the chromium layer. The in general more intensive mold cleaning is the critical step and limits mainly the lifetime of HNMs. The degradation of the chromium layer can be compensated by initially thicker chromium layers. However, this limits the achievable resolution during the mold manufacturing process. A recoating of the HNM could also solve the problem of the chromium degradation. A challenge for this recoating process is the required high resolution. A process for recoating would be also beneficial for polymer molds, which are not based on the standard hard mask processes during production. The development of such a process was not intended for this work.

It was found that antisticking layers degraded fast on antireflective chromium surfaces (cf. Sec. 4.1.3) indicating a different behavior of ASLs on metal oxide compared to silica surfaces. However, the chromium oxide on HNMs cannot be compared with the antireflective chromium used for those experiments due to the modification of the antireflective surface during mold etching. Under optimized CNP conditions, the residual layer is not cured and the adhesion to the chromium is low. Chromium oxide behaves only as a light lewis acid compared to silica (cf. Tab. 2.3). Thus, the adhesion to basic resist systems like acrylates is expected to be low compared to silica.

As already mentioned, due to the removal of parts of the the anti-reflective hard mask (Hoya Corp.) during the fused silica etching, the original layer was modified. No data on this new layer could be obtained and the transmittance had to be estimated. A 50 nm thick chromium layer has a transmittance of roughly 0.2 % including fresnel loss for normal incidence[8]. For a 20 nm chromium layer the transmittance is increased to about 4 %. The initial 70 nm anti-reflective chromium hard mask had an optical density of 3.1 at 248 nm and 2.2 at blue light[9]. Assuming an optical denity of 3 for 365 nm and using the direct proportionality of the optical density for thinned films, the 50 nm and a 20 nm layer had an optical density of approximately 2 and 1 or a transmittance of 1 % and 10 %, respectively. In conclusion, the tranmittance of the 50 nm layer was roughly below 1 %. The transmittance of the much thinner chromium was about 4-10 %.

5.4.3 Residual Layer Curing

Whether the resist under the chromium was cured or not depended in addition to the used dose and the transmitted intensity on the resist itself and the used photoinitiator. A comparable low-sensitiv initiator was used in the customized, acrylate based resist formulation (see Fig. 5.25b). This customized resist gave a better contrast between chromium covered and uncovered regions than UVCur21SF. With the same chromium layer, the dispensed customized resist under the chromium was not cured upon exposure with 6 J/cm^2. In contrast, spin-coated UVCur21SF was already cured through the chromium layer at doses below 2 J/cm^2. Residual layer free imprints were more difficult to achieve with the higher sensitivity and the lower contrast of UVCur21SF. When UVCur21SF was pre-exposed, the required imprint doses had to be larger than the accumulated doses. For

[8] n=1.861+2.681i, R=0.51, air-chromium interface (www.refractiveindex.info)
[9] according to Hoya Corp.

example the residual layer of an imprint site with an accumulated dose of 4.8 J/cm^2 was not cured upon exposure with 4.8 J/cm^2 but was completely cured after being exposed with 9 J/cm^2. Pre-exposure reduces the photoinitiator concentration, which is clearly responsible for successful curing. Thus, the curing behavior under the chromium can be influenced by choosing an appropriate photoinitiator and its concentration. A consequence for spin-coated UVCur21SF films was, that only for the first SR-UV-NIL imprint, the residual layer will be cured. Subsequent fields accumulated enough dose to prevent a curing of the residual layer for the standard imprint dose 1.96 J/cm^2.

Especially for regions with high feature densities, the residual layer was partially cured in the surrounding feature area even though the residual layer further apart from the features was not cured (see Fig. 5.28a). Close to the features the critical dose was probably reached as due to indirect light propagation in the resist and light transmission through the chromium. Transmission through the chromium alone cannot be the reason because residual layer regions in large distance to the features were not cured. A dome like RLT-profile similar to completely cured residual layer (cf. Fig. 5.22) was found for the exposed residual layer close to features (see Fig. 5.28b).

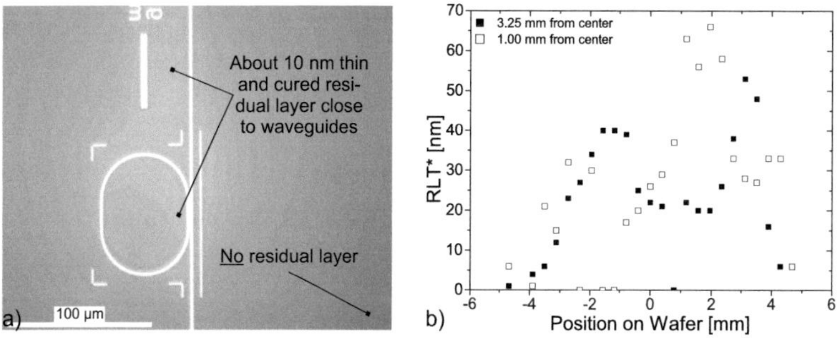

Figure 5.28: *a) Thin, cured residual layer due to light reflection, diffraction and scattering into the residual layer gap under the chromium cover and transmission through the chromium as observed already on a 90 nm SiO_2-layer and b) dome-like shape of the RLT-inhomogeneity of partially exposed residual layers at different positions in the imprint field (*: Measured close to structures, did not represent the fully cured RLT between mold and substrate).*

However, the thickness of the residual layer was only about 66 nm close to the center of the imprint field. Closer to the periphery, the residual layer in the exposed areas was much thinner. Visually, the residual layer was also not completely cured to its full thickness. The curing started obviously from the substrate-resist interface for the imprints on 90 nm thick thermally grown silicon oxide. Assumed the dome-like shape of the residual layer was present under the mold, the resist filled gap between mold and substrate was in the center larger than at the mold periphery. Additionally to the tinner layers, also the exposed area around the structures was smaller at the periphery. This motivates the idea, that a thickness dependent light transport existed. Light reflection, refraction, scattering and diffraction were expected to be the reason for a light propagation into the residual layer gap. A simple geometric optics model was used to estimate the intensity distribution under the chromium layer (see Fig. 5.29).

As discussed in Sec. 5.1, the waveguides required a low refractive under-cladding typically in the order of 10 µm. The available materials are transparent at 365 nm UV-light used to cure the imprint materials. The used UV-LED source had a broad angular spectrum (see Fig. 5.30a). It was postulated for this model, that the UV-LED intensity distribution as it leaves the imprint exposure system was not modified by the mold and was the same at the openings in the chromium layer. The refractive index of the used resist UVCur21SF and SiO_2 was 1.535 [161] and 1.475 [153]b, respectively. The customized acrylate resists had refractive indices similar to the UVCur21SF and the resulting intensity distributions are thus comparable. The critical angle for the path resist-SiO_2 was thus $\theta_c \approx 74°$. Angles larger than this value accounted for less than 1 % of the UV-angular intensity distribution (cf. Fig 5.30a). Light with those angles of incident were totally reflected at the resist-SiO_2 interface. Light with smaller angles of incident was also refracted into the SiO_2. Thereby the angle of refraction $\theta_{r,2}$ and thus the angle of incidence on the silicon substrate was increased equally. An incident angle $\theta_{i,1}$=74° was translated to an angle of reflectance of $\theta_{r,2}$=85°. The reflection on the Si-interface occurs at the same angle $\theta_{r,2}$. Silicon has a refractive index of 6.547 at a wavelength of 365 nm [153]d. The reflection coefficient of non-polarized light for incident angles between 0 and 74° is about 0.5 including dielectric losses at the silicon interface (cf. Figs. in Apendix C). The resist and the SiO_2 were modeled without dielectric losses.

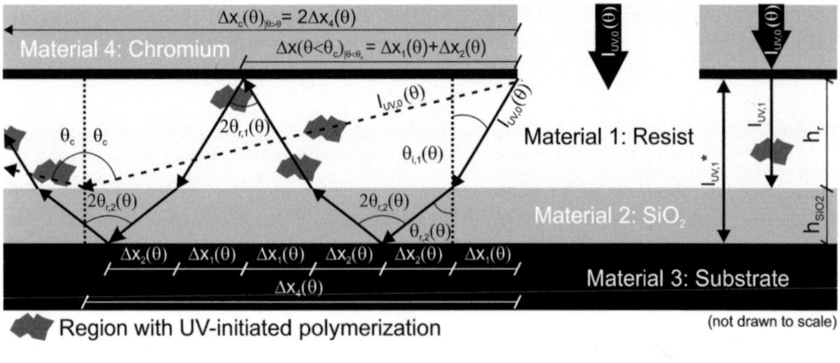

Figure 5.29: *Simple geometric optics model for light propagation into a residual polymer layer.*

For the calculations, the amplitude coefficient of reflection were calculated from the complex refractive indices, squared and multiplied with the respective relative intensity to calculate the lateral intensity distribution. Fig. 5.30b shows this lateral intensity distribution along the chromium surface before reflection on it. This distribution considered not the resist close to the resist-SiO_2 interface exposed directly due to the broad angular intensity distribution. Light with an incidence angle of 55°carrying still about 50 % of the UV-LED peak intensity can directly reach only as far as 300 nm beyond the chromium edge. The highest intensity at the chromium is observed directly at the cavity edge with less than 25% of the peak intensity. In about 1μm distance from the chromium edge the intensity dropped to less than 10% of the peak intensity. About 0.5-1.0 mJ/cm^2 were estimated to be the lower limit required for at least a partial curing of the residual layer (cf. Sec. 5.2.2). Such high value were only reached in the very narrow region close to the imprinted features. Additionally, reflection with dielectric loss occurs at the resist-chromium interface. However, the intensity was already too low to consider further multiple reflections.

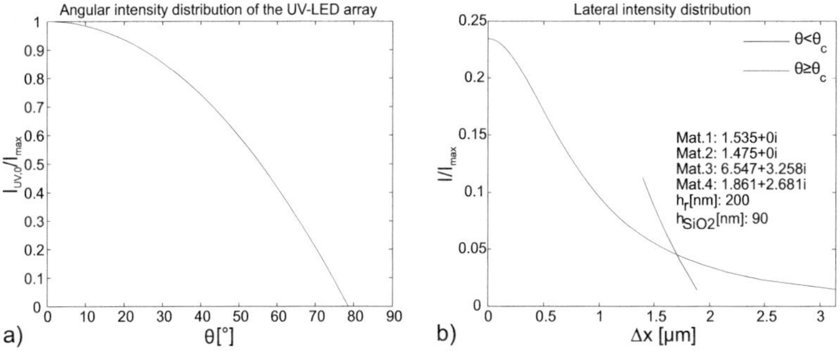

Figure 5.30: *Intensity distribution of the used UV-LED array approximated with a quadratic function [according to the manufacturer of the NPS300 imprint system (SET S.A.S)] and b) estimated lateral intensity distribution based on the geometric optics model in Fig. 5.29 (reflection / refraction).*

In conclusion the estimated narrow region of less than 500 nm around the imprint features alone cannot explain the observed large areas of partially cured residual layers. Only in sum with the transmitted light as well as effects like scattering leading to a more diffuse light distribution, the partial curing of the residual layer can be explained. Large areas of the MRR-layout can be kept residual layer free. The gap of the MRR cannot be kept residual layer free if the dose is not optimized and the sensitivity and the contrast of the resist is suitable. However, a residual layer free gap is possible. In general, especially sub-wavelength features will suffer from diffraction. It is more difficult to replicate residual layer free on substrates with transparent top layers like SiO_2.

6 Conclusion and Outlook

"If your experiment needs statistics, then you ought to have done a better experiment."

(Ernest Rutherford)

6.1 Conclusion

This work defines the term direct polymer patterning for techniques which directly use polymer patterns as functional parts of a device and not as mask for further pattern transfer. The scope of this work was to evaluate the applicability of UV-assisted nanoimprint lithography (UV-NIL) for the emerging field of direct polymer patterning. A photonic microring resonator (MRR) intended for a biosensor was chosen as technology demonstrator. The major reason for using UV-NIL was the potential to pattern application-specifically synthesized imprint polymers. Additionally, very-low viscous materials were required for imprint molds with low pattern densities. First, basic technology components were established. Based on these components, a technology for a medium-volume production of MRRs was proposed. Based on the experience made with the MRR, the suitability of UV-NIL for direct polymer patterning was rated.

A chemical vapor deposition (CVD) system for silanes was developed during this work. This system enabled high quality anti-sticking layers and a structure conformal coating of high aspect ratio molds. Compared to published data on antisticking layer coatings, the stability of the vapor coated layers were comparable high for multiple UV-NIL imprints. However, the commonly observed anti-sticking layer degradation was also observed for the antisticking layer used in this work. The maximum amount of possible imprints was in the order of a few hundreds. Metal oxides were proposed as alternative, non-degrading antisticking layers.

Customized resist synthesis is already used in research, e.g., to decrease the anti-sticking layer degradation by the incorporation of fluorine. In this work, a hydroxyl-functional imprint resist was synthesized. The accessibility of the superficial hydroxyl-groups on unpatterned polymer films was demonstrated with chemical derivatization with a fluorine tag. This approach demonstrated, that application specifically functionalized imprint polymers are an important part of future NIL techniques.

To reduce imprint mold costs, especially during the development and testing of new materials and imprint parameters, the known concept of bonded imprint molds was improved. Different adhesives were proposed for permanently as well as for demountable bonded molds. In a new approach, the chemical stability of these adhesives against acidic piranha, a common mold cleaning agent, could be significantly improved by vapor coating of the adhesives with trichlorosilanes in the before mentioned silane CVD-process.

A major contribution towards the universal application of silicon as imprint mold material for UV-NIL was made with a new technique termed non-transparent UV-NIL (NT-UV-NIL). With this technique, opaque molds were imprinted by UV-NIL on opaque substrates. The advantage of this technique is that standard UV-NIL imprint systems with an exposure unit located above the mold can be used without modification. In combination with metal-oxide antisticking layers or opaque polymeric molds with intrinsic anti-sticking properties, this technique is a solution for antisticking layer degradation in UV-NIL. Completely new materials can be used as mold materials. The only requirement is a transparent layer of currently about 20 µm on an opaque substrate. NT-UV-NIL is not suited for the pattern transfer to the opaque substrate but appropriate for lift-off patterning and direct polymer nanoimprinting.

The basic components were combined into a proposal for a medium volume MRR-production technology. Attention was also put on post-imprint processes like automatic wafer handling and final chip-dicing, but the main focus was on the imprinting process. In contrast to the established drop-on-demand dispensing of optimized resist patterns, this work proposes imprinting in spin-coated films. Imprints on spin-coated films had significantly higher yields due to much less defects due to trapped air and self-assembling. Spin-coated films require further much less optimization efforts than dispensed patterns.

This work contributed to the field of stamp-and-repeat imprinting in spin-coated films by demonstrating the effects of the pre-exposure on subsequent imprints if no light blocking can be done: The pre-exposure reduced the amount of available photoinitiator molecules in the subsequent imprint fields, which resulted in curing defects after a few imprints. This maximum allowed accumulated dose was given by the half maximal effective dose D_{50}. This value was obtained from a symmetrical, sigmoidal dose-response-curve of the demolding force F_d plotted against the accumulated dose. F_d was a reliable and easy to access measure for the degree of curing of the imprint polymer.

The imprint process was described with the impulse and could be easily derived from the imprint force-time-profile. Comparable large impulses, meaning forces equivalent to about 100 N and holding times of about 5 min, were required to flow the polymer into the cavities of the highly negative mold. Filling from thin initial resist films was very challenging. Capillary filling of the used highly negative molds with micron-sized cavities was not observed. This was mainly because the imprints were done under ambient conditions without evacuation of the imprint zone and due to the large cavities.

The residual layer thickness obtained with the highly negative imprint mold was quite inhomogeneous with about 250 nm in the center and less than 5 nm at the periphery of a $10\times10\,mm^2$ large imprint field. The source of this inhomogeneity was attributed to an inhomogeneous hydrostatic pressure resulting in a mold and substrate deformation which further hampers the imprint material flow towards the mold periphery. The inhomogeneous hydrostatic pressure originated from the mold layout and comparable thick initial layers for this layout. A modified mold layout with dummy-fill structures adapted to a film thickness of about 500 nm would help to avoid the deformation.

Another solution to avoid the inhomogeneity was the use of hybrid imprint molds (HNMs). These molds have a light blocking metal layer on top of the imprint protrusions and were manufactured by leaving the chromium hard-mask from fused silica etching on the imprint mold. HNMs avoided the complete curing of the residual layer. Depending on the resist sensitivity to UV-light, the residual layer could be avoided completely. A challenge for sub-wavelength features and imprinting on semi- or completely transparent substrates are diffuse light components and diffraction. Both led to a partial exposure of the residual layer.

The proposed manufacturing technology demonstrated the challenges of nanoimprinting as purely mechanical patterning technique. Replicating certain mold structures is less a challenge than achieving reproducible results with high yields. Nanoimprint lithography is highly layout dependent. This work contributed mainly to UV-based nanoimprinting techniques as a dedicated technique for polymer direct patterning as a future industrial NIL-approach.

6.2 Outlook

A perspective MRR-biosensor will require the quantification of the available hydroxyl-groups or other functional groups after imprinting. The antisticking layer material itself might be an additional possibility to tune the functional density on the patterned polymer surface. The application of NT-UV-NIL as flexible polymer patterning technique depends on the influence of the layout as well as the feature size and height on the curing resist behavior. This work should inspire further research in these fields.

To minimize the pre-exposure during stamp-and-repeat UV-NIL, light blocking coatings or apertures are required. If the coatings are applied to the imprint mold, they have to be stable against typical imprint processes. The success of apertures will depend on their integration into imprint systems without influencing the imprint quality and the thermal load of the exposure system.

The existing technology approach can be extended to three dimensional and multi-tier structures. For these structures the shrinkage behavior will become a major aspect of a successful pattern transfer and an interesting area of further research.

Appendices

A Model for the Fowkes Dispersion Approach

This approach is based on two assumption made by *Fowkes*. First he assumed that the surface energy can be devided into different components stemming from the different intermolecular forces discussed in Section 2.1 (d...dispersion, p...polar dipole-dipole, i...polar dipole-induced dipole). As stated by *Pocius* [23] this is only an approximation because potential energies cannot be summed up without accounting for cross-relations, which is not done in this approximation.

$$\gamma = \gamma^d + \gamma^p + \gamma^i + ... \tag{A.1}$$

Fowkes stated further that a apolar liquid, which has only a γ^d contribution, can only interact with the γ^d contribution of the solid in contact with the liquid. This layed the ground for what is today known as theory of *Fractional Polarity* saying bascially that only like forces interact with each other. For a completely apolar liquid interacting with a solid Φ becomes unity. This can be phenomenologically explained by the fact that Fowkes' hypothesis states that the apolar liquid behaves in the same way like the solids bulk apolar forces. Equation (2.11) becomes

$$\gamma_{12} = \gamma_1 + \gamma_2 - 2\sqrt{\gamma_1^d \gamma_2^d}. \tag{A.2}$$

Putting this equation into the Dupré relation (2.5) gives

$$Wa = \gamma_1 + \gamma_2 - (\gamma_1 + \gamma_2 - 2\sqrt{\gamma_1^d \gamma_2^d}) = 2\sqrt{\gamma_1^d \gamma_2^d}. \tag{A.3}$$

This relates W_a directly to the known surface tension of a apolar liquid and allows the calculation of the dispersion component of the unknown solid with equation (2.9). Combining those equations give (with the correct assumption $\gamma_{LV}= \gamma_{LV}^d$ for apolar liquids.

$$\gamma_{LV}(1 + cos\Theta) + \pi = 2\sqrt{\gamma_{LV}^d \gamma_{SV}^d} \tag{A.4}$$

For low energy surface and high tension liquids ($\pi = 0$), γ_{SV}^d could be directly calculated.

B Derivation of the relation of the initiator concentration I on the amount of subsequent exposures n

According to *Dickey* [147], the time dependent photoinitiator concentration I can be modeled as

$$\frac{dI}{dt} = -\Phi' \cdot k_I \cdot I = -c \cdot I \tag{B.1}$$

Φ' is the number of initiator molecules dissociated per absorbed photon and solely depends on the photoinitiator itself. The coefficient of initiation k_I depends solely on the amount of absorbed photons. For convenience, both parameters are assumed to be constant over multiple exposures and are combined in the constant c.

The function of I over the time t is denoted as $f(t)$. To get the amount of initiator molecules after a short time step Δt of exposure, f is evolved as linear taylor series at the point ($f(a)$,a) and termed $T_1(t)_{|a}$:

$$T_1(t)_{|a} = f(a) + f'(a)(t-a) \tag{B.2}$$

For the first exposure starting at t=t_0, the concentration of the initiator I=$f(t_0)$ is known. This point is used to evolve the taylor series. After the exposure, a certain value I_{t_1} is reached at t_1=t_0+Δt. This second point can be expressed with the above derived taylor series for small Δt and a=t_0.

$$I_{t_1} = I_1 = T_1(t_1)_{|t_0} = f(t_0) + f'(t_0)(t_1 - t_0) \tag{B.3}$$

This initial point ($f(t_0)$,t_0) is given by $f(t_0)$=I_0 and $f'(t)$=$-c \cdot I_0$ and the above stated equation can be written as

$$I_1 = I_0 - c \cdot I_0 \cdot (t_1 - t0) = I_0 - c \cdot I_0 \cdot \Delta t = I_0(1 - c \cdot \Delta t) \tag{B.4}$$

For the next exposure, starting at the now known point ($f(t_1)$,t_1), the same taylor series can be used to express the point ($f(t_2)$,t_2) with t_2=t_1+Δt=t_0+$2\Delta t$.

$$I_{t_2} = I_2 = I_1 \cdot (1 - c \cdot \Delta t) = I_0 \cdot (1 - c \cdot \Delta t)^2 \tag{B.5}$$

The initiator concentration I_n after n exposures each for a time step Δt can be expressed using the initial concentration I_0 and the relation

$$I_{t_0 + n\Delta t} = I_n = I_0 \cdot (1 - c \cdot \Delta t)^n = \tau^n \cdot I_0. \tag{B.6}$$

C Coefficients of Specular Reflection for the Interfaces between UVCur21SF, SiO2, Si and Chromium at λ=365 nm

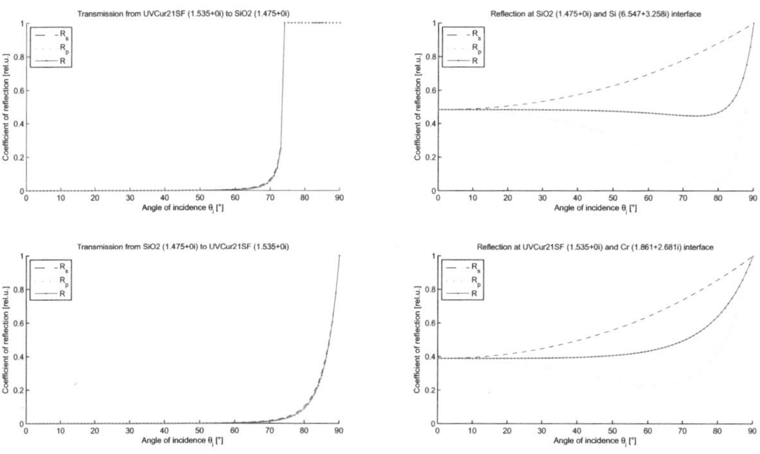

Figure C.1: *Coefficients of reflection for different interfaces as they occur in a typical UV-NIL setup after passing the mask opening of an HNM.*

List of Publications

Parts of the work done during this PhD thesis were published as following.

Journals

1. R. Kirchner, L. Teng, B. Lu, B. Adolphi, and W.-J. Fischer, "Degradation of Perfluorotrichlorosilane Antisticking Layers: The Impact on Mold Cleaning, UV-Nanoimprinting, and Bonded UV-Nanoimprint Molds," *Jpn. J. Appl. Phys.*, vol. 50, p. 06GK13 (8pp), 2011

2. R. Kirchner, A. Finn, L. Teng, M. Ploetner, A. Jahn, L. Nueske, and W.-J. Fischer, "UV-Nanoimprinting Using Non-Transparent Molds and Non-Transparent Substrates," *Microelectron. Eng., vol. 88, pp. 2004-2008, 2011.*

3. R. Kirchner, M.-K. Kaiser, B. Adolphi, R. Landgraf, and W.-J. Fischer, "Chemical functional polymers for direct UV assisted nanoimprinting of polymeric photonic microring resonators," *Phys. Status Solidi A*, vol. 208, no. 6, pp. 1308-1314, 2011

4. R. Kirchner, B. Adolphi, R. Landgraf, and W.-J. Fischer, "Antisticking Layers on Antireflective Chromium for Hybrid (CNP) Nanoimprint Molds," *Proc. SPIE*, vol. 7545, pp. 7545-0U, 2010

Proceedings

5. R. Landgraf, A. Finn, R. Kirchner, W.-J. Fischer, F. Deicke, "Biosensor System With Polymer Microring Resonators Manufactured by Multilevel-UV-Nanoimprint," to be published in *Proc. Mikro-System-Technik Kongress 2011*, 2011 [accepted]

6. R. Landgraf, T. Haugwitz, R. Kirchner, A. Finn, W.-J. Fischer, "Planar Optical Waveguide Design for UV-Nanoimprinted Microring Resonator Based Biosensors," to be published in *Proc. IEEE Sensors 2011*, 2011 [accepted]

7. R. Kirchner, L. Nueske, A. Finn, B. Lu, W.-J. Fischer, "Stamp-and-Repeat UV-Imprinting of Spin-Coated Films: Pre-Exposure and Imprint Defects," to be published in *Proc. 37th Int. Conf. on Micro and Nano Engineering 2011 (MNE 2011)*, 2011 [accepted]

8. L. Teng, R. Kirchner, M. Ploetner, J. He, A. Jahn, F. Hagemann, W.-J. Fischer, "Fabrication and Characterization of Sub-500 nm Channel Organic Field Effect

Transistor using UV Nanoimprint Lithography with opaque molds," to be published in *Proc. 37th Int. Conf. on Micro and Nano Engineering 2011 (MNE 2011)*, 2011 [accepted]

9. R. Kirchner, A. Kleiner, B. Adolphi, W.-J. Fischer, "Comparison of silane monolayers on fused silica and chromium-oxi-nitride surfaces by angle resolved photoelectron spectroscopy and a new multi-layer model", in *Proc. Engineering of Functional Interfaces*, (Linz), p. 23-24, 19.-20.07.2011

10. L. Teng, R. Kirchner, M. Ploetner, A. Jahn, J. He, F. Hagemann, and W.-J. Fischer, "Fabrication of Sub-500 nm Source and Drain Electrodes for Organic Field Effect Transistors using UV Nanoimprint Lithography with Low-Cost Silicon Mold and Lift-Off Process," in *3rd GMM Workshop Mikro-Nano-Integration*, vol. GMM Fachbericht 68, pp. 84-89, VDE Verlag GmbH, 3th-4th March 2011. ISBN 978-3-8007-3334-7 [in English]

11. A. Finn, A. Jahn, R. Kirchner, U. Kuenzelmann, J. He, M. Waegner, and W.-J. Fischer, "Multilayer Nano-Imprint-Lithography Mold Fabrication Process," in *3rd GMM Workshop Mikro-Nano-Integration*, vol. GMM Fachbericht 68, pp. 8-12, VDE Verlag GmbH, 3th-4th March 2011. ISBN 978-3-8007-3334-7 [in English]

12. R. Kirchner, L. Teng, B. Lu, B. Adolphi, and W.-J. Fischer, "Degradation of Perfluorotrichlorosilane Antisticking Layers and the Impact on Bonded UV-Nanoimprint Molds, Mold Cleaning, and UV-Nanoimprinting,"' in *Proc. 23rd Internat. Microprocesses and Nanotechnology Conf.*, (Kokura / Fukuoka), 09.-12.11.2010 2010. 12D-11-99

13. R. Kirchner, R. Kullock, L. M. Eng, and W.-J. Fischer, "Residual Layer Free Polymer Waveguides and Microring Resonators Realized with UV-assisted Nanoimprint," in *Proc. 9th Internat. Conf. on Nanoimprint and Nanoprint Technology*, pp. 88-89, 2010

14. R. Kirchner, A. Finn, L. Teng, M. Ploetner, A. Jahn, L. Nueske, and W.-J. Fischer, "UV-Nanoimprinting Using Non-Transparent Molds and Non-Transparent Substrates," in *Proc. 36th Int. Conf. on Micro and Nano Engineering 2010 (MNE 2010)*, 2010. P-LITH-92

15. R. Kirchner, M.-K. Kaiser, B. Adolphi, R. Landgraf, and W.-J. Fischer, "Chemical Functional Polymers for Direct UV Assisted Nanoimprinting of a Polymeric Photonic Biosensor," in *Proc. Engineering of Functional Interfaces*, (Rauischholzhausen / Marburg), p. 59, 15.-16.07.2010. Paper 3.6

16. R. Kirchner, R. Landgraf, M. Bertram, and W.-J. Fischer, "Direct UV-nanoimprint of polymer microring resonators as optical transducers," in *Proc. 2nd Workshop Mikro-Nano-Integration*, pp. 153-158, VDE Verlag GmbH, 2010. ISBN: 978-3-8007-3216-6 [in English]

17. R. Kirchner, L. Teng, and W.-J. Fischer, "Multi-usable, adhesively bonded UV-NIL templates," in *Proc. 8th Internat. Conf. on Nanoimprint and Nanoprint Technology*, p. p52, 2009

18. R. Kirchner, M.-K. Kaiser, B. Adolphi, R. Landgraf, and W.-J. Fischer, "Imprinting of chemical functionalized acrylates for bioanalytics," in *Proc. 6th nanomed*, (Berlin), p. 127, 04.-06.03.2009 2009. ISBN 978-3-00-027401-5

Conferences and Workshops (without proceedings)

19. R. Kirchner, L. Teng, and W.-J. Fischer, "Bonded nanoimprint templates using perflourinated trichlorosilane monolayers as protection layers during template cleaning," in presented at *Mikro-System-Technik Kongress 2009*, (Berlin), p. Poster 11.8, 12.-14.10.2009 2009. [late paper, not in proceedings, available upon request]

20. R. Landgraf, R. Kirchner, and W.-J. Fischer, "Photonic Biosensors based on Microring Resonators - High Integration and Low-Cost Sensing," in presented at *Korean-German-Symposium on Medical Diagnostic*, (Dresden), 29.10.2009 2009

21. R. Kirchner, B. Adolphi, K. Richter, and W.-J. Fischer, "Reduced PDMS swelling in toluene and acrylates by C4F8 plasma fluorination," in *Max Bergmann Symposium*, (Dresden), 2008. [as supplemental sheet, available upon request]

22. R. Landgraf, A. Naumann, R. Kirchner, and W.-J. Fischer, "Photonic Biosensing based on Microring-Resonators," in *Max Bergmann Symposium*, (Desden), 2008. [as supplemental sheet, available upon request]

Talks and Lectures

23. R. Kirchner and A. Türke, "Using Ink-Jet Printing and Nanoimprinting for Microsystems," in presented at Industrie-Partner-Symposium TU Dresden at Semicon Europe, (Dresden), 06.10.2010

24. R. Kirchner, "Nanoimprinting - Basics and Applications in Microsystems Technology," in presented at *3. Colloquium Halbleitertechnologie / Mikroelektronik - Strukturierung in Mikroelektronik und Mikrosystemtechnik*, (Dresden), Silicon Saxony e. V., 08.06.2009

25. R. Kirchner, "Nanoimprinting for high aspect ratio optical microsystems," in presented at *Industrie-Partner-Symposium TU Dresden at Semicon Europe*, (Dresden), 06.10.2009

Patents

26. R. Kirchner and W.-J. Fischer, "Verfahren und Vorrichtung zur Herstellung von Nanoimprinttemplates," 2010. DE 10.2010.043.235 A1

27. R. Kirchner, M. Ploetner, and W.-J. Fischer, "Imprinttemplate, Nanoimprintvorrichtung und Nanostrukturierungsverfahren," 2010. DE 10.2010.043.059 A1

28. R. Kirchner, M. Ploetner, and W.-J. Fischer, "Imprint Template, Nanoimprint Device and Nanostructuring Method," 2010. US 2011/0115130 A1

Bibliography

[1] ITRS, "Lithography," in *International Technology Roadmap for Semiconductors (2009 Edition)*, 2009.

[2] S. Y. Chou, P. R. Krauss, and P. J. Renstrom, "Imprint of sub-25 nm vias and trenches in polymers," *Appl. Phys. Lett.*, vol. 67, no. 21, pp. 3114–3116, 1995.

[3] C.-Y. Chao, W. Fung, and L. J. Guo, "Polymer Microring Resonators for Biochemical Sensing Applications," *IEEE J. Quantum. Elect.*, vol. 12, pp. 134–142, 2006.

[4] J. Haisma, M. Verheijen, K. van den Heuvel, and J. van den Berg, "Mold-assisted nanolithography: A process for reliable pattern replication," *J. Vac. Sci. Technol. B*, vol. 14, no. 6, pp. 4124–4128, 1996.

[5] M. Colburn, S. C. Johnson, M. D. Stewart, S. Damle, T. C. Bailey, B. Choi, M. Wedlake, T. B. Michaelson, S. V. Sreenivasan, J. G. Ekerdt, and C. G. Willson, "Step and flash imprint lithography: a new approach to high-resolution patterning," *Proc. SPIE*, vol. 3676, pp. 379–389, 1999.

[6] ITRS, "Lithography," in *International Technology Roadmap for Semiconductors (2003 Edition)*, 2003.

[7] H. Cao, Z. Yu, J. Wang, J. O. Tegenfeldt, R. H. Austin, E. Chen, W. Wu, and S. Y. Chou, "Fabrication of 10 nm enclosed nanofluidic channels," *Appl. Phys. Lett.*, vol. 81, no. 1, pp. 174–176, 2002.

[8] L. J. Guo, X. Cheng, and C. Y. Chao, "Fabrication of Photonic Nanostructures in Nonlinear Optical Polymers," *J. Mod. Optic.*, vol. 49, no. 3, pp. 663–673, 2002.

[9] T. Balla, S. M. Spearing, and A. Monk, "An assessment of the process capabilities of nanoimprint lithography," *Journal of Physics D: Applied Physics*, vol. 41, no. 17, p. 174001 (10pp), 2008.

[10] D. Moser, M. Heinrich, C. Schuster, A. Klukowska, and A. Schmidt, "Development and Characterization of a Process for Microlense Fabrication by Lithography, RIE and UV molding," in *MikroSystemTechnik Kongress 2009*, 2009.

[11] H. Schmitt, M. Rommel, A. J. Bauer, L. Frey, A. Bich, M. Eisner, R. Voelkel, and M. Hornung, "Full wafer microlens replication by UV imprint lithography," *Microelectron. Eng.*, vol. 87, no. 5-8, pp. 1074–1076, 2010.

[12] C. Peroz, S. Dhuey, A. Goltsov, M. Volger, B. Harteneck, I. Ivonin, A. Bugrov, S. Cabrini, S. Babin, and V. Yankov, "Digital spectrometer-on-chip fabricated by step andrepeat nanoimprint lithography on pre-spin coated films," *Microelectron. Eng.*, p. 4pp, 2011. [article in press].

[13] J. Taniguchi, Y. Kamiya, T. Ohsaki, and N. Sakai, "Technique for transfer of high-density, high-aspect-ratio nanoscale patterns in UV nanoimprint lithography and measurement of the release force," *Microelectron. Eng.*, vol. 87, pp. 859–863, 2010.

[14] G. Kumar, H. X. Tang, and J. Schroers, "Nanomoulding with amorphous metals," *Nature (letters)*, vol. 457, pp. 868–873, 2009.

[15] European Technology Platform Photonics21, "Photonics - Our Vision for a Key Enabling Technology in Europe." [Preliminary edition, February 2011].

[16] H. Ma, A. K.-Y. Jen, and L. R. Dalton, "Polymer-Based Optical Waveguides: Material, Processing, and Devices," *Adv. Mater.*, vol. 14, no. 19, pp. 1339–1365, 2002.

[17] C.-Y. Chao, *Polymer Microring Resonator and its Application as a Biosensor*. PhD thesis, University of Michigan, 2005.

[18] J. T. Kim and C.-G. Choi, "Fabrication of Roughness-Improved Polymer Waveguides by UV Imprint Techniques," *Jpn. J. Appl. Phys.*, vol. 44, no. 5A, pp. 3073–3074, 2005.

[19] T. Bailey, B. J. Choi, M. Colburn, M. Meissl, S. Shaya, J. G. Ekerdt, S. V. Sreenivasan, and C. G. Willson, "Step and flash imprint lithography: Template surface treatment and defect analysis," *J. Vac. Sci. Technol. B*, vol. 18, no. 6, pp. 3572–3577, 2000.

[20] R. Ji, M. Hornung, M. Verschuuren, R. van de Laar, J. van Eekelen, U. Plachetka, M. Moeller, and C. Moormann, "UV enhanced substrate conformal imprint lithography (UV-SCIL) technique for photonic crystals patterning in LED manufacturing," *Microelectron. Eng.*, vol. 87, no. 5-8, pp. 963–967, 2009.

[21] R. Landgraf, "Photonic Viral Diagnostics by Using Microring Resonators (working title)." unpublished PhD work, 2012.

[22] W. C. Wake, *Adhesion and the Formulation of Adhesives*. Applied Science Publishers Ldt., 1976. ISBN 0-85334-660-7.

[23] A. V. Pocius, *Adhesion and Adhesives Technology - An Introduction*. München, Wien: Hanser, 2nd ed., 2002.

[24] H.-C. Scheer, H. Schulz, T. Hoffmann, and C. M. Sotomayor-Torres, *Handbook of Thin Film Materials*, vol. Nanomaterials and Magnetic Thin Films, ch. Nanoimprint Techniques, p. 49. Academic Press, 2002.

[25] W. Kern, ed., *Handbook of Semiconductor Wafer Cleaning Technology.* Park Ridge: Noyes Publications, 1993.

[26] L.-H. Lee, ed., *Fundamentals of Adhesion.* New York and London: Plenum Press, 1991.

[27] E. McCafferty, "Acid-base effects in polymer adhesion at metal surfaces," *J. Adhesion Sci. Technol.*, vol. 16, no. 3, pp. 239–255, 2002.

[28] B. Sellergren, ed., *Molecularly imprinted polymers: Man-made mimics of antibodies and their applications in analytical chemistry.* Amsterdam-Lausanne-New York-Oxford-Shannon-Singapore-Tokyo: Elsevier, 2001.

[29] K. L. Mittal and H. R. Anderson, eds., *Acid-Base Interactions: Relevance to Adhesion Science and Technology.* Utrecht: VSP, 1991.

[30] E. McCafferty, "Lewis Acid/Lewis Base Effects in Corrosion and Polymer Adhesion at Aluminum Surfaces," *J. Electrochem. Soc.*, vol. 150, no. 7, pp. B342–B347, 2003.

[31] S. Lee, J.-S. Park, and T. R. Lee, "The Wettability of Fluoropolymer Surfaces: Influence of Surface Dipoles," *Langmuir*, vol. 24, no. 9, pp. 4817–4826, 2008.

[32] E. M. Harnett, J. Alderman, and T. Wood, "The surface energy of various biomaterials coated with adhesion molecules used in cell culture," *Colloids Surf. B: Biointerfaces*, vol. 55, pp. 90–97, 2006.

[33] E. G. Shafrin and W. A. Zisman, "Constitutive Relations in the Wetting of Low Energy Surfaces and the Theory of the Retraction Method of Preparing Monolayers," *J. Phys. Chem.*, vol. 64, pp. 519–524, May 1960.

[34] S. Wu, *Polymer Handbook*, ch. VI - Solid State Properties. John Wiley & Sons, Inc., 4th ed., 1999.

[35] R. W. Jaszewski, H. Schift, B. Schnyder, A. Schneuwly, and P. Groening, "The deposition of anti-adhesive ultra-thin teflon-like films and their interaction with polymers during hot embossing," *Appl. Surf. Sci.*, vol. 143, pp. 301–308, 1999.

[36] D.-Y. Khang, H. Kang, T.-I. Kim, and H. H. Lee, "Low-Pressure Nanoimprint Lithography," *Nano Lett.*, vol. 4, no. 4, pp. 633–637, 2004.

[37] D.-Y. Khang and H. H. Lee, "Sub-100 nm Patterning with an Amorphous Fluoropolymer Mold," *Langmuir*, vol. 20, pp. 2445–2448, Mar. 2004.

[38] J. P. Rolland, R. M. V. Dam, D. A. Schorzman, S. R. Quake, and J. M. DeSimone, "Solvent-Resistant Photocurable "Liquid Teflon" for Microfluidic Device Fabrication," *J. Am. Chem. Soc.*, vol. 126, pp. 2322–2323, 2004.

[39] D. R. Barbero, M. S. M. Saifullah, P. Hoffmann, H. J. Mathieu, D. Anderson, G. A. C. Jones, M. E. Welland, and U. Steiner, "High Resolution Nanoimprinting with a Robust and Reusable Polymer Mold," *Adv. Funct. Mater.*, vol. 17, pp. 2419–2425, 2007.

[40] M. Verschuuren and H. van Sprang, "3D Photonic Structures by Sol-Gel Imprint Lithography," in *Mater. Res. Soc. Symp. Proc. Vol. 1002*, pp. 1002–N03–05, 2007.

[41] S. Gilles, M. Meier, M. Proempers, A. van der Hart, C. Kuegeler, A. Offenhaeusser, and D. Mayer, "UV nanoimprint lithography with rigid polymer molds," *Microelectron. Eng.*, vol. 86, pp. 661–664, 2009.

[42] A. Cattoni, E. Cambril, D. Decanini, G. Faini, and A. Haghiri-Gosnet, "Soft UV-NIL at 20 nm scale using flexible bi-layer stamp casted on HSQ master mold," *Microelectron. Eng.*, vol. 86, pp. 586–589, 2009.

[43] M. Muehlberger, I. Bergmair, A. Klukowska, A. Kolander, H. Leichtfried, E. Platzgummer, H. Loeschner, C. Ebm, G. Gruetzner, and R. Schoeftner, "UV-NIL with working stamps made from Ormostamp," *Microelectron. Eng.*, vol. 86, pp. 691–693, 2009.

[44] N. Kehagias, V. Reboud, J. D. Girolamo, M. Chouiki, M. Zelsmann, J. Boussey, and C. S. Torres, "Stamp replication for thermal and UV nanoimprint lithography using a UV-sensitive silsesquioxane resist," *Microelectron. Eng.*, vol. 86, pp. 776–778, 2009.

[45] Y. Xia and G. Whitesides, "Soft lithography," *Annual Review of Materials Science*, vol. 28, no. 1, pp. 153–184, 1998.

[46] J. N. Lee, C. Park, and G. M. Whitesides, "Solvent Compatibility of Poly(dimethylsiloxane)-Based Microfluidic Devices," *Anal. Chem.*, vol. 75, pp. 6544–6554, 2003.

[47] H. Schmid and B. Michel, "Siloxane Polymers for High-Resolution, High-Accuracy Soft Lithography," *Macromolecules*, vol. 22, pp. 3042–3049, 2000.

[48] L. Montelius, B. Heidari, M. Graczyk, I. Maximov, E.-L. Sarwe, and T. Ling, "Nanoimprint- and UV-lithography: Mix&Match process for fabrication of interdigitated nanobiosensors," *Microelectron. Eng.*, vol. 53, no. 1-4, pp. 521–524, 2000.

[49] G. M. McClelland, M. W. Hart, C. T. Rettner, M. E. Best, K. R. Carter, and B. D. Terris, "Nanoscale patterning of magnetic islands by imprint lithography using a flexible mold," *Appl. Phys. Lett.*, vol. 81, no. 8, pp. 1483–1485, 2002.

[50] F. A. Houle, C. T. Rettner, D. C. Miller, and R. Sooriyakumaran, "Antiadhesion considerations for UV nanoimprint lithography," *Appl. Phys. Lett.*, vol. 90, p. 213103, 2007.

[51] K. ichiro Nakamatsu, N. Yamada, K. Kanda, Y. Haruyama, and S. Matsui, "Fluorinated Diamond-Like Carbon Coating as Antisticking Layer on Nanoimprint Mold," *Jpn. J. Appl. Phys.*, vol. 45, no. 35, pp. L954–L956, 2006.

[52] D. Janssen, *Self-assembling Monolayers for Organic Thin-film Transistors.* PhD thesis, Katholieke Universiteit Leuven, 2006.

[53] T. M. Mayer, M. P. de Boer, N. D. Shinn, P. J. Clews, and T. A. Michalske, "Chemical vapor deposition of fluoroalkylsilane monolayer films for adhesion control in microelectromechanical systems," *J. Vac. Sci. Technol. B*, vol. 18, no. 5, pp. 2433–2440, 2000.

[54] M. J. Stevens, "Thoughts on the Structure of Alkylsilane Monolayers," *Langmuir*, vol. 15, pp. 2773–2778, 1999.

[55] B. C. Bunker, R. W. C. abd R. A. Assink, M. L. Thomas, M. G. Hankins, J. A. Voigt, D. Sipola, M. P. de Boer, and G. L. Gulley, "The Impact of Solution Agglomeration on the Deposition of Self-Assembled Monolayers," *Langmuir*, vol. 16, pp. 7742–7751, 2000.

[56] G.-Y. Jung, Z. Li, W. Wu, Y. Chen, D. L. Olynick, S.-Y. Wang, W. M. Tong, and R. S. Williams, "Vapor-Phase Self-Assembled Monolayer for Improved Mold Release in Nanoimprint Lithography," *Langmuir*, vol. 21, no. 4, pp. 1158–1161, 2005.

[57] D. Truffier-Boutry, R. Galand, A. Beaurain, A. Francone, B. Pelissier, M. Zelsmann, and J. Boussey, "Mold cleaning and fluorinated anti-sticking treatments in nanoimprint lithography," *Microelectron. Eng.*, vol. 86, pp. 669–672, 2009.

[58] "The physical and theoretical chemistry laboratory - oxford university. chemical and other safety information: Safety data for 1h,1h,2h,2h-perfluorodecyltrichlorosilane." online. accessed 29.03.2011.

[59] D. Briggs and M. P. Seah, eds., *Practical Surface Analysis, Volume I - Auger and x-ray photoelectron spectroscopy.* Chichester: John Wiley and Sons, 1990.

[60] R. W. Paynter, "Angle-resolved x-ray photoelectron spectroscopy studies of the evolution of plasma-treated surfaces with time," *Surf. Interface Anal.*, vol. 27, no. 2, pp. 103–113, 1999.

[61] M. Kozlowska, R. Reiche, S. Oswald, H. Vinzelberg, R. Hübner, and K. Wetzig, "Quantitative ARXPS investigation of systems with ultrathin aluminium oxide layers," *Surf. Interface Anal.*, vol. 36, no. 13, pp. 1600–1608, 2004.

[62] M. Kozlowska, *Winkelaufgelöste XPS: Optimierung der mathematischen Modellierung und deren experimentelle Überprüfung.* PhD thesis, Technische Universität Dresden, 2004. [in German].

[63] A. Kleiner, "Herstellung, Charaktersierung und Vergleich von Antihaftschichten auf Si-, SiO2- und Chromoberflächen für das UV-Nanoimprinting," Master's thesis, Technische Universität Dresden, Institute of Semiconductor and Microsystems Technology, 2010. [in German].

[64] C. C. Perry, A. J. Wagner, and D. H. Fairbrother, "Electron stimulated C-F bond breaking kinetics in fluorine-containing organic thin films," *Microelectron. Eng.*, vol. 280, no. 1-2, pp. 111–118, 2002.

[65] P. J. Cumpson and M. P. Seah, "Elastic Scattering Corrections in AES and XPS. II. Estimating Attenuation Lengths and Conditions Required for their Valid Use in Overlayer/Substrate Experiments," *Surface and Interface Analysis*, vol. 25, no. 6, pp. 430–446, 1997.

[66] D. Y. Fozdar, W. Zhang, M. Palard, C. W. Patrick(Jr), and S. Chen, "Flash imprint lithography using a mask aligner: a method for printing nanostructures in photosensitive hydrogels," *Nanotechnology*, vol. 19, p. 215303, 2008.

[67] S. R. Wasserman, Y.-T. Tao, and G. M. Whitesides, "Structure and Reactivity of Alkylsiloxane Monolayers Formed by Reaction of Alkyltrichlorosilanes on Silicon Substrates," *Langmuir*, vol. 5, pp. 1074–1087, 1989.

[68] N. Yoshino, Y. Kondo, and T. Yamauchi, "Syntheses and reactions of metal organics. XXL Syntheses of (1H,1H,2H,2H-polyfluoroalkyl)triisocyanate silanes and surface modification of glass," *J. Fluorine Chem.*, vol. 79, no. 1, pp. 87–91, 1996.

[69] J. Taniguchi, T. Kawasaki, Y. Tokano, Y. Kogo, I. Miyamoto, M. Komuro, H. Hiroshima, N. Sakai, and K. Tada, "Measurement of Adhesive Force Between Mold and Photocurable Resin in Imprint Technology," *Jpn. J. Appl. Phys.*, vol. 41, pp. 4194–4197, 2002.

[70] S. Garidel, M. Zelsmann, P. Voisin, N. Rochat, and P. Michallon, "Structure and stability characterization of anti-adhesion selfassembled monolayers formed by vapour deposition for NIL use," *Proc. SPIE*, vol. 6517, p. 65172C, 2007.

[71] H. Schmitt, M. Zeidler, M. Rommel, A. J. Bauer, and H. Ryssel, "Custom-specific UV nanoimprint templates and life-time of antisticking layers," *Microelectron. Eng.*, vol. 85, no. 5-6, pp. 897–901, 2008.

[72] D. Truffier-Boutry, M. Zelsmann, J. D. Girolamo, J. Boussey, C. Lombard, and B. Pépin-Donat, "Chemical degradation of fluorinated antisticking treatments in UV nanoimprint lithography," *Appl. Phys. Lett.*, vol. 94, no. 4, p. 044110, 2009.

[73] M. Zelsmann, C. Alleaume, D. Truffier-Boutry, A. Francone, A. Beaurain, B. Pelissier, and J. Boussey, "Degradation and surfactant-aided regeneration of fluorinated anti-sticking mold treatments in UV nanoimprint lithography," *Microelectron. Eng.*, vol. 87, pp. 1029–1032, 2010.

[74] F. A. Houle, E. Guyer, D. C. Miller, and R. Dauskardt, "Adhesion between template materials and UV-cured nanoimprint resists," *J. Vac. Sci. Technol. B*, vol. 25, no. 4, pp. 1179–1185, 2007.

[75] K. Kuwabara, A. Miyauchi, and H. Sugimura, "Degradation behavior of release layers for nanoimprint lithography formed on atomically flat Si(111) terraces," *J. Vac. Sci. Technol. B 28*, vol. 28, no. 5, pp. 968–972, 2010.

[76] J. Y. Kim, D.-G. Choi, J.-H. Jeong, and E.-S. Lee, "UV-curable nanoimprint resin with enhanced anti-sticking property," *Appl. Surf. Sci.*, vol. 254, pp. 4793–4796, 2008.

[77] H. Hiroshima, "Release force reduction in UV nanoimprint by mold orientation control and by gas environment," *J. Vac. Sci. Technol. B*, vol. 27, no. 6, pp. 2862–2865, 2009.

[78] G. Gerlach, W. Dötzel, J. Mehner, and N.-T. Nguyen, *Einführung in die Mikrosystemtechnik.* Fachbuchverlag Leipzig im Carl Hanser Verlag, 2006. [in German].

[79] H. J. Levinson, ed., *Principles of lithography.* SPIE-The International Society for Optical Engineering, 2nd ed., 2005.

[80] S. Inoue, K. Kajikawa, and Y. Aoyagi, "Dry-etching method for fabricating photonic-crystal waveguides in nonlinear-optical polymers," *Appl. Phys. Lett.*, vol. 82, pp. 2966–2968, 2003.

[81] S. J. Pearton and D. P. Norton, "Dry Etching of Electronic Oxides, Polymers, and Semiconductors," *Plasma Processes Polym.*, vol. 2, pp. 16–37, 2005.

[82] M. A. McCord and M. J. Rooks, *SPIE Handbook of Microlithography, Micromachining and Microfabrication*, ch. 2: Electron Beam Lithography, pp. 139–232. SPIE-The International Society for Optical Engineering, 1997.

[83] S. Juodkazis, V. Mizeikis, and H. Misawa, "Three-Dimensional Structuring of Resists and Resins by Direct Laser Writing and Holographic Recording," *Adv. Polym. Sci.*, vol. 213, pp. 157–206, 2008.

[84] C. M. S. Torres, ed., *Alternative Lithography - Unleashing the Potentials of Nanotechnology.* Kluwer Academic / Plenum Publishers, 2003.

[85] L. J. Guo, "Recent progress in nanoimprint technology and its applications," *Journal of Physics D: Applied Physics*, vol. 37, no. 11, pp. R123–R141, 2004.

[86] L. J. Guo, "Nanoimprint Lithography: Methods and Material Requirements," *Adv. Mater.*, vol. 19, pp. 495–513, 2007.

[87] B. Bhushan, ed., *Springer Handbook of Nanotechnology.* Springer, 2007.

[88] H. Schift, "Nanoimprint lithography: An old story in modern times? A review," *J. Vac. Sci. Technol. B*, vol. 26, no. 2, pp. 458–480, 2008.

[89] H.-C. Scheer, N. Bogdanski, M. Wissen, and S. Möllenbeck, "Impact of glass temperature for thermal nanoimprint," *J. Vac. Sci. Technol. B*, vol. 25, no. 6, pp. 2392–2395, 2007.

[90] H.-C. Scheer and H. Schulz, "A contribution to the flow behaviour of thin polymer films during hot embossing lithography," *Microelectron. Eng.*, vol. 56, pp. 311–332, 2001.

[91] T. Hoffmann, *Alternative Lithography: Unleashing the Potentials of Nanotechnology*, ch. Viscoelastic Properties of Polymers, pp. 25–45. Kluwer Academic / Plenum Publishers, 2003.

[92] F. Xu, N. A. Stacey, M. Watts, V. Truskett, I. McMackin, J. Choi, P. Schumaker, E. Thompson, D. Babbs, S. V. Sreenivasan, C. G. Willson, and N. Schumaker, "Development of imprint materials for the Step and Flash Imprint Lithography process," *Proc. SPIE*, vol. 5374, p. 232, 2004.

[93] X. Cheng and L. J. Guo, "One-step lithography for various size patterns with a hybrid mask-mold," *Microelectron. Eng.*, vol. 71, no. 3-4, pp. 288 – 293, 2004.

[94] L. H. D. Skjolding, G. T. Teixidor, J. Emneus, and L. Montelius, "Negative UV-NIL (NUV-NIL) - A mix-and-match NIL and UV strategy for realisation of nano- and micrometre structures," *Microelectron. Eng.*, vol. 86, pp. 654–656, 2009.

[95] H.-C. Scheer, M. Wissen, N. Bogdanski, S. Möllenbeck, and A. Mayer, "Potential and limitations of a T-NIL/UVL hybrid process," *Microelectron. Eng.*, vol. 87, pp. 851–853, 2010.

[96] S. Moellenbeck, N. Bogdanski, A. Mayer, and H.-C. Scheer, "Positive or negative tone resist for a T-NIL/UVL hybrid process," *Proc. SPIE*, vol. 7545, pp. 75450T–1, 2010.

[97] M. Wissen, H. Schulz, N. Bogdanski, H. C. Scheer, Y. Hirai, H. Kikuta, G. Ahrens, F. Reuther, and K. Pfeiffer, "UV curing of resists for warm embossing," *Microelectron. Eng.*, vol. 73-74, pp. 184–189, 2004.

[98] X. Cheng and L. J. Guo, "A combined-nanoimprint-and-photolithography patterning technique," *Microelectron. Eng.*, vol. 71, no. 3-4, pp. 277–282, 2004.

[99] K. Okuda, N. Niimi, H. Kawata, and Y. Hirai, "Micro-nano mixture patterning by thermal-UV novel nanoimprint," *J. Vac. Sci. Technol. B*, vol. 25, pp. 1859–1861, 2007.

[100] Y. Chen, F. Carcenac, C. Ecoffet, D. J. Lougnot, and H. Launois, "Mold-assisted near-field optical lithography," *Microelectron. Eng.*, vol. 46, pp. 69–72, 1999.

[101] Y. Hirai, H. Kikuta, and T. Sanou, "Study on optical intensity distribution in photocuring nanoimprint lithography," *J. Vac. Sci. Technol. B*, vol. 21, no. 6, pp. 2777–2782, 2003.

[102] H. Schift and L. J. Heydermann, *Alternative Lithography - Unleashing the Potentials of Nanotechnology*, ch. 4. Nanorheology, pp. 47–76. Kluwer Academic / Plenum Publishers, 2003.

[103] T. Haugwitz, "Entwurf und Charakterisierung planaroptischer Mikroresonatoren," Master's thesis, Fachhochschule Mittweida, Mathematik-Naturwissenschaften-Informatik, 2011.

[104] X. Liang, H. Tan, Z. Fu, and S. Y. Chou, "Air bubble formation and dissolution in dispensing nanoimprint lithography," *Nanotechnology*, vol. 18, p. 025303 (7pp), 2007.

[105] M. Hoga, K. Itoh, M. Ishikawa, N. Kuwahara, M. Fukudaa, N. Toyama, S. Kurokawa, and T. Doi, "Comparison of quartz and silicon as a master mold substrate for patterned media UV-NIL replica process," *Microelectron. Eng.*, 2011. [in press].

[106] C.-Y. Chao and L. J. Guo, "Reduction of Surface Scattering Loss in Polymer Microrings Using Thermal-Reflow Technique," *IEEE Photonic. Tech. L.*, vol. 16, no. 6, pp. 1498–1500, 2004.

[107] B. Wu, A. Kumar, and S. Pamarthy, "High aspect ratio silicon etch: A review," *J. Appl. Phys.*, vol. 108, pp. 051101 (1–20), 2010.

[108] H.-C. Liu, Y.-H. Lin, and W. Hsu, "Sidewall roughness control in advanced silicon etch process," *Microsyst. Technol.*, vol. 10, pp. 29–34, 2003.

[109] K. J. Morton, G. Nieberg, S. Bai, and S. Y. Chou, "Wafer-scale patterning of sub-40 nm diameter and high aspect ratio (>50:1) silicon pillar arrays by nanoimprint and etching," *Nanotechnology*, vol. 19, p. 345301, 2008.

[110] D. Nilsson, S. Jensen, and A. Menon, "Fabrication of silicon molds for polymer optics," *J. Micromech. Microeng.*, vol. 13, pp. S57–S61, 2003.

[111] M. V. Bazylenko and M. Gross, "Reactive ion etching of silica structures for integrated optics applications," *J. Vac. Sci. Technol. A*, vol. 14, no. 6, pp. 2994–3003, 1996.

[112] Q. Xia, P. F. Murphy, H. Gao, and S. Y. Chou, "Ultrafast and selective reduction of sidewall roughness in silicon waveguides using self-perfection by liquefaction," *Nanotechnology*, vol. 20, p. 345302, 2009.

[113] S. K. Bag, D. S. Mainardi, and N. P. Mahalik, *Features of Surface Micromachining and Wafer Bonding Process*, ch. 6, pp. 107–130. Springer Berlin Heidelberg, 2006.

[114] M. Chiao and L. Lin, *MEMS Packaging*, ch. Ch3 - Sealing Technologies, pp. 61–84. INSPEC, 2004.

[115] A. Plössl and G. Kräuter, "Wafer direct bonding: tailoring adhesion between brittle materials," *Mater. Sci. Eng., R*, vol. 25, no. 1-2, pp. 1–88, 1999.

[116] J. Lee, A. Ali, K. don Kim, J.-H. Kim, D. guen Choi, J.-H. Choi, and J. ho Jeong, "Plasma-assisted quartz-to-quartz direct bonding for the fabrication of a multilayered quartz template for nanoimprint lithography ," *J. Micromech. Microeng.*, vol. 20, no. 4, p. 045005, 2010.

[117] H. Schmitt, *Untersuchung der UV-Nanoimprint-Lithografie als Strukturierungsverfahren für elektronische Bauelemente*. PhD thesis, Universität Erlangen-Nürnberg, 2008. [in German].

[118] J. M. Goddard and J. H. Hotchkiss, "Polymer surface modification for the attachment of bioactive compounds," *Prog. Polym. Sci.*, vol. 32, pp. 698–725, 2007.

[119] M. Jaeger, A. Meyer-Plath, F. Sonntag, J. Friedrich, and R. Poll, "Selective amino functionalization of polymer surfaces: UV-laser irradiation compared to plasma processing," in *18th Int. Symposium on Plasma Chemistry*, (Kyoto), p. 423, 2007.

[120] Y. Ofir, I. W. Moran, C. Subramani, K. R. Carter, and V. M. Rotello, "Nanoimprint Lithography for Functional Three-Dimensional Patterns," *Adv. Mater.*, vol. 22, pp. 3608–3614, 2010.

[121] M. Behl, J. Seekamp, S. Zankovych, C. Torres, R. Zentel, and J. Ahopelto, "Towards Plastic Electronics: Patterning Semiconducting Polymers by Nanoimprint Lithography," *Adv. Mater.*, vol. 14, p. 588, 2002.

[122] M. B. Chan-Park, Y. Yan, W. K. Neo, W. Zhou, J. Zhang, and C. Y. Yue, "Fabrication of High Aspect Ratio Poly(ethylene glycol)-Containing Microstructures by UV Embossing," *Langmuir*, vol. 19, pp. 4371–4380, 2003.

[123] I. Park, J. Cheng, A. P. Pisano, E.-S. Lee, and J.-H. Jeong, "Low temperature, low pressure nanoimprinting of chitosan as a biomateria for bionanotechnology applicationsl," *Appl. Phys. Lett.*, vol. 90, p. 093902, 2007.

[124] H. Schmitt, F. Kett, M. Rommel, A. J. Bauer, M. Hornung, and L. Frey, "Direct imprinting, post processing, and characterization of functional UV-Curing Materials," in *Proc. 9th Internat. Conf. on Nanoimprint and Nanoprint Technology*, 2010.

[125] J. Choi, S. Jung, D. Choi, J. Jeong, and E. Lee, "Direct soft UV-NIL with resist incorporating carbon nanotubes," *Microelectron. Eng.*, vol. 85, pp. 195–201, 2008.

[126] B. K. Long, B. K. Keitz, and C. G. Willson, "Materials for step and flash imprint lithography (S-FIL)," *J. Mater. Chem.*, vol. 17, pp. 3575–3580, 2007.

[127] K. P. Toralla, J. D. Girolamo, D. Truffier-Boutry, and M. Zelsmann, "High flowability monomer resists for thermal nanoimprint lithography," *Microelectron. Eng.*., vol. 86, pp. 779–782, 2009.

[128] D. Morihara, H. Hiroshima, and Y. Hirai, "Numerical study on bubble trapping in UV-nanoimprint lithography," *Microelectron. Eng.*, vol. 86, pp. 684–687, 2009.

[129] T. Lee, C. Guymonc, E. S. Jönsson, and C. Hoyle, "The effect of monomer structure on oxygen inhibition of (meth)acrylates photopolymerization," *Polymer*, vol. 45, pp. 6155–6162, 2004.

[130] N. Tham, "Praktikumsbeleg: Entwicklung einer Methode zur Antihaftbeschichtung von Nanoimprintstempeln sowie einer Technologie zur Herstellung einer Flusszelle für die Biosensorik," tech. rep., Fraunhofer Institut für Photonische Mikrosysteme (IPMS), 2009. [in German].

[131] H. Schift, S. Saxer, S. Park, C. Padeste, U. Pieles, and J. Gobrecht, "Controlled co-evaporation of silanes for nanoimprint stamps," *Nanotechnology*, vol. 16, no. 5, pp. S171–S175, 2005.

[132] B. Lu, "Studienarbeit: Herstellung und Charakterisierung von silanbasierten Antihaftschichten für Nanoimprinting," tech. rep., Technische Universität Dresden, Institute of Semiconductor and Microsystems Technology, 2010. [in German].

[133] I. M. Tidswell, B. M. Ocko, P. S. Pershan, S. R. Wasserman, G. M. Whitesides, and J. D. Axe, "X-ray syecular reSection studies of silicon coated by organic mono]ayers (alkylsiloxanes)," *Phys. Rev. B.*, vol. 41, no. 2, pp. 1111–1128, 1990.

[134] M. Cichomski, K. Kosla, J. Grobelny, W. Kozlowski, J. Kowalczyk, A. Busiakiewicz, W. Szmaja, and J. Balcerski, "Nano- and microtribological characterization of silanes deposited on cobalt substrate," *J. Alloy. Compd.*, vol. 507, no. 1, pp. 273–278, 2010.

[135] G. L. Witucki, "A Silane Primer: Chemistry and Applications of AIkoxy Silanes," *J. Coat. Technol.*, vol. 65, no. 822, pp. 57–60, 1993.

[136] S. Garidel, M. Zelsmann, N. Chaix, P. Voisin, J. Boussey, A. Beaurain, and B. Pelissier, "Improved release strategy for UV nanoimprint lithography," *J. Vac. Sci. Technol. B*, vol. 25, no. 6, pp. 2430–2434, 2007.

[137] L. Teng, "Untersuchung der Anwendbarkeit sowie Charakterisierung der UV basierten Nanoimprinttechnik mit nichttransparenten Stempeln anhand der Polymerdirektstrukturierung und des Lift-Off-Prozesses," Master's thesis, Technische Universität Dresden, Institute of Semiconductor and Microsystems Technology, 2010. [in German].

[138] *Datasheet Ciba DAROCUR 1173 (Edition: 4.9.2001, Basle).*

[139] J. G. A. Terlingen, H. F. C. Gerritsen, A. S. Hoffman, and J. Feijen, "Introduction of Functional Groups on Polyethylene Surfaces by a Carbon Dioxide Plasma Treatment," *J. Appl. Polym. Sci.*, vol. 57, pp. 969–982, 1995.

[140] L. Teng, "Herstellung und Charakterisierung hybrider Stempelaufbauten für das Nanoimprinting," student project, Technische Universität Dresden, Institute of Semiconductor and Microsystems Technology, 2009. [in German].

[141] I. Epoxy Technology, *EPO-TEK 305 Technical Data Sheet*, 08 2007. [Rev. IV - 08/2007].

[142] DELO Industrie Klebstoffe, *DELO Technische Information: DELO-Katiobond 4670*. [01.10 (Revison 26)].

[143] L. Teng, R. Kirchner, M. Ploetner, A. Jahn, J. He, F. Hagemann, and W.-J. Fischer, "Fabrication of Sub-500 nm Source and Drain Electrodes for Organic Field Effect Transistors using UV Nanoimprint Lithography with Low-Cost Silicon Mold and Lift-Off Process," in *3rd GMM Workshop Mikro-Nano-Integration*, vol. GMM Fachbericht 68, pp. 84–89, VDE Verlag GmbH, 3th-4th March 2011. ISBN 978-3-8007-3334-7 [in English].

[144] H. S. Lee, D. S. Kim, and T. H. Kwon, "UV nano embossing for polymer nano structures with non-transparent mold insert," *Microsyst. Technol.*, vol. 13, no. 5, pp. 593–599, 2007.

[145] C. Decker and D. Decker, "Photoinitiated radical polymerization of vinyl ether-maleate systems," *Polymer*, vol. 38, no. 9, pp. 2229–2237, 1097.

[146] L. Lecamp, B. Youssef, and C. Bunel, "Photoinitiated polymerization of a dimethacrylate oligomer: 1. Influence of photoinitiator concentration, temperature and light intensity," *Polymer*, vol. 38, no. 25, pp. 6089–6096, 1997.

[147] M. D. Dickey and C. G. Willson, "Kinetic parameters for step and flash imprint lithography photopolymerization," *AIChE J.*, vol. 52, pp. 777–784, 2006.

[148] H. Schmitt, L. Frey, and H. Ryssel, "UV nanoimprint materials: Surface energies, residual layers, and imprint quality," *J. Vac. Sci. Technol. B*, vol. 25, no. 3, pp. 785–790, 2007.

[149] R. Landgraf, T. Haugwitz, R. Kirchner, A. Finn, and W.-J. Fischer, "Planar Optical Waveguide Design for UV-Nanoimprinted Microring Resonator Based Biosensors," in *Proc. IEEE Sensors 2011*, pp. 596–599, 2011.

[150] J.-Q. Xi, M. F. Schubert, J. K. Kim, E. F. Schubert, S.-Y. Chen, Minfeng Lin, W. Liu, and J. A. Smart, "Optical thin-film materials with low refractive index for broadband elimination of Fresnel reflection," *Nature Photonics*, vol. 1, pp. 176–179, 2007.

[151] Honeywell Electronic Materials, *Thin Films-Dielectrics*, 2004.

[152] Sigma-Aldrich, "Low refractive index materials." online, 06 2011. updated 01.06.2011.

[153] M. Polyanskiy, "Refractive index database." online on 28.05.2011, 05 2011. [a) Handbook of Optics, 3rd edition, Vol. 4. McGraw-Hill 2009 and M. J. Dodge, Refractive properties of magnesium fluoride, Appl. Opt. 23, (1984), p. 1980-1985; b) Fused silica, I. H. Malitson. Interspecimen Comparison of the Refractive Index of Fused Silica, J. Opt. Soc. Am. 55 (1965), p. 1205-1208 and Handbook of Optics, 3rd edition, Vol. 4. McGraw-Hill 2009; c) extrapolated from T. Bååk. Silicon oxynitride; a material for GRIN optics, Appl. Optics 21 (1982), p. 1069-1072 ; d) crystalline, Handbook of Optical Constants of Solids, Edward D. Palik. Academic Press, Boston, 1985; e) extrapolated from S. N. Kasarova et al., Analysis of the dispersion of optical plastic materials, Optical Materials 29 (2007), p. 1481-1490].

[154] micro resist technology GmbH, *UV-härtende Polymere für die UV-gestützte Nanoimprint-Lithographie: mr-UVCur21 und mr-UVCur06 - Schnellhärtende Polymersysteme zur Strukturübertragung*. 08.09.25.01.

[155] J. W. Jeong, H. H. Ryoo, J. M. Hong, S. G. Lee, B. H. O, and E. H. Lee, "Photodefinable Polyimide as an Effective New Material for Passive Photonic Devices," *J. Korean Phys. Soc.*, vol. 39, no. 2, pp. 250–254, 2001.

[156] A. Borreman, S. Musa, A. A. M. Kok, M. B. J. Diemeer, and A. Driessen, "Fabrication of Polymeric Multimode Waveguides and Devices in SU-8 Photoresist Using Selective Polymerization," in *Proc. Symposium IEEE/LEOS Benelux Chapter*, (Amsterdam), 2002.

[157] micro resist technology GmbH, *Processing guidelines: OrmoCore*. 9113002.

[158] micro resist technologie GmbH, *Personal correspondence*, 12 2010.

[159] micro resist technology GmbH, *Processing guidelines: mr-I T85 - Thermoplastic Polymer for Nanoimprint Lithography*. 08040203.

[160] W.-J. Liu, Y.-T. Lai, Y.-C. Lai, M.-H. Weng, and C.-M. Chen, "Rib-Type Polyimides and Polyimide-Silica Hybrid Arrayed Waveguide Grating Simulation and Fabrication," *Proc. SPIE*, vol. 5723, pp. 297–306, 2005.

[161] micro resist technology GmbH, *Processing guidelines: mr-UVCur21SF - UV-curable Polymer for UV-based Nanoimprint Lithography*. 08090501.

[162] micro resist technology GmbH, *Processing guidelines: mr-UVCur06 - UV-curable Polymer System for UV-based Nanoimprint Lithography*. 09113005.

[163] T. C. Bailey, M. Colburn, B. J. Choi, A. Grot, J. G. Ekerdt, S. V. Sreenivasan, and C. G. Willson, *Alternative Lithography - Unleashing the Potentials of Nanotechnology*, ch. 7. Step and Flash Imprint Lithography, pp. 117–137. Kluwer Academic / Plenum Publishers, 2003.

[164] C. Peroz, S. Dhuey, M. Vogler, Y. Wu, D. Olynick, and S. Cabrini, "Step and repeat UV nanoimprint lithography on pre-spin coated resist film: a promising route for fabricating nanodevices," *Nanotechnology*, vol. 21, p. 445301 (5pp), 2010.

[165] H. Hiroshima and M. Komuro, "Control of Bubble Defects in UV-Nanoimprint," *Jpn. J. Appl. Phys.*, vol. 46, no. 9B, pp. 6391–6394, 2007.

[166] H. Schift, L. J. Heyderman, M. A. D. Mauer, and J. Gobrecht, "Pattern formation in hot embossing of thin polymer films," *Nanotechnology*, vol. 12, pp. 173–177, 2001.

[167] N. Chaix, C. Gourgon, S. Landis, C. Perret, M. Fink, F. Reuther, and D. Mecerreyes, "Influence of the molecular weight and imprint conditions on the formation of capillary bridges in nanoimprint lithography," *Nanotechnology*, vol. 17, pp. 4082–4987, 2006.

[168] N. Chaix, S. Landis, D. Hermelin, T. Leveder, C. Perret, V. Delaye, and C. Gourgon, "Influence of mold depth and capillary bridges in nanoimprint lithography," *J. Vac. Sci. Technol. B*, vol. 24, no. 6, pp. 3011–3015, 2006.

[169] H. Atasoy, M. Vogler, T. Haatainen, A. Schleunitz, H. Schift, F. Reuther, and G. Gruetzner, "Reduction of the Release Forces of Polymers for Thermal NIL and their Quantification," in *Proc. 9th Internat. Conf. on Nanoimprint and Nanoprint Technology*, 2010.

[170] D. Y. Fozdar, W. Zhang, M. Palard, C. W. J. Patrick, and S. Chen, "Flash imprint lithography using a mask aligner: a method for printing nanostructures in photosensitive hydrogels," *Nanotechnology 19*, vol. 19, p. 215303 (13pp), 2008.

[171] H.-C. Scheer, Y. Hirai, N. Bogdanski, and M. Nishihata, *Polymer elasticity effects during thermal nanoimprint.* Der Andere Verlag, 2008.

[172] S. GmbH, *Applications: Free Radical Initiators.* [downloaded 29.04.2008].